Resilience, Authenticity, and Digital Heritage Tourism

This book examines the authentication of authenticity in heritage tourism by using a resilient smart systems approach. It discusses the emerging trends in cultural tourism and outlines, in a detailed manner, their significance in negotiating the authenticity of tourism experiences.

Authentication of authenticity is an evolving, less-researched field of inquiry in heritage tourism. This book advances research on this subject by exploring different authentication processes and scrutinizes their resilience in building transformative heritage tourism pathways. It offers a kaleidoscopic view of the manner authenticity has evolved over the last several decades by observing a broad spectrum of cultural expressions. The evolution and meaningfulness of negotiated authenticity is identified and discussed in the context of pre-, intra-, and post-pandemic times. This book focuses on the moral and existentialist trajectories of authenticity and the notion of self-authentication. It proposes a smart resilient authentication model to delicately negotiate the objective and self-dimensions of authenticity in transformative times. Furthermore, by sharing examples of best practices, it offers unique insights on how authenticity is authenticated and mediated via digital platforms and artificial intelligence.

This book offers novel perspectives on negotiated authenticity and its authentication in heritage tourism and will appeal to both practitioners and students/scholars of Heritage studies; Design and Innovation; Tourism Studies; Geography and Planning across North America, Europe, and East Asian countries.

Deepak Chhabra is an associate professor and senior sustainability scientist at Arizona State University, United States. Her research interests include authenticity and authentication of heritage, economic equity, smart/sustainable marketing strategies for heritage tourism, and alternative healing/restorative systems and eudaimonic well-being of both visited and visiting communities.

Routledge Cultural Heritage and Tourism Series
Series editor: Dallen J. Timothy, Arizona State University, USA

The Routledge Cultural Heritage and Tourism Series offers an interdisciplinary social science forum for original, innovative and cutting-edge research about all aspects of cultural heritage-based tourism. This series encourages new and theoretical perspectives and showcases ground-breaking work that reflects the dynamism and vibrancy of heritage, tourism and cultural studies. It aims to foster discussions about both tangible and intangible heritages, and all of their management, conservation, interpretation, political, conflict, consumption and identity challenges, opportunities and implications. This series interprets heritage broadly and caters to the needs of upper-level students, academic researchers, and policy makers.

The Economics and Finance of Cultural Heritage
How to Make Tourist Attractions a Regional Economic Resource
Vincenzo Pacelli and Edgardo Sica

Urban Recovery
Intersecting Displacement with Post War Reconstruction
Edited by Howayda Al-Harithy

Resilience, Authenticity and Digital Heritage Tourism
Deepak Chhabra

Cultural Heritage and Tourism in Japan
Takamitsu Jimura

Medieval Imaginaries in Tourism, Heritage and the Media
Jennifer Frost and Warwick Frost

For more information about this series, please visit: https://www.routledge.com/Routledge-Cultural-Heritage-and-Tourism-Series/book-series/RCHT

Resilience, Authenticity, and Digital Heritage Tourism

Deepak Chhabra

Routledge
Taylor & Francis Group
LONDON AND NEW YORK

First published 2022
by Routledge
2 Park Square, Milton Park, Abingdon, Oxon OX14 4RN

and by Routledge
605 Third Avenue, New York, NY 10158

Routledge is an imprint of the Taylor & Francis Group, an informa business

British Library Cataloguing-in-Publication Data
A catalogue record for this book is available from the British Library

Library of Congress Cataloging-in-Publication Data
A catalog record has been requested for this book

ISBN: 978-0-367-56656-2 (hbk)
ISBN: 978-0-367-56663-0 (pbk)
ISBN: 978-1-003-09883-6 (ebk)

Typeset in Times
by KnowledgeWorks Global Ltd.

Contents

Tables

Figures

Preface

From the time I joined my PhD program at North Carolina State University (Raleigh, USA), I have kept track of ongoing deliberations on the authenticity phenomenon. My dissertation, for the most part, had focused on the authenticity of heritage festivals (Chhabra 2001). During the past nineteen years, I have advanced my research in this field by exploring, identifying, and testing new settings and pathways. Additionally, I have viewed a variety of authentication processes associated with numerous heritage expressions and settings. In my book titled '*Sustainable Marketing of Cultural and Heritage Tourism*', I featured authenticity as one of the core elements in the proposed marketing model and critically scrutinized it, using a sustainability perspective while contextualizing it within numerous heritage expressions. Undeniably, documented literature is saturated with numerous journal articles on authenticity. However, books examining negotiated authenticity and authentication of heritage, from a resilience viewpoint, are sparse. To my knowledge, no book to date has scrutinized digital authentication of heritage in an in-depth manner. Pressing focus on innovative practices from negotiated authentication and digital as well as digital detox standpoints makes this book a useful contribution in the pandemic times; it also opens the path for further deliberations, particularly in the context of heritage resilience and sustainability.

Authenticity, undoubtedly, continues to be the cornerstone of heritage tourism and is now considered a strategic component in managing sustainable heritage tourism. United Nations designated 2017 as the 'year of sustainable tourism for development'. Authenticity of heritage tourism is featured as an important contributor to this theme. It is of no surprise then that the authenticity discourse has taken center stage during the past several decades and has been the subject for various deliberations and analyses. It has become a new path to accomplish smart development of heritage tourism (Lugosi 2016; Park, Choi & Lee 2019; Pine & Gilmore 2008). Multiple views have shaped the discursive path of authenticity, with special attention to essentialist/object (genuine, true to the origin), constructivist (commodified for revenue purpose), existentialist (seeking optimized and euphoric state of mind), and negotiated authenticities centered on both supply and demand perspectives (Cohen 2002; Kolar & Zabkar 2010; MacCannell 1976; Pine & Gilmore 2008; Reisinger & Steiner 2006; Wang 1999).

As stated by MacCannell, a cultural production can serve one of the two vital roles: "it may add to the weight of the modern civilization by sanctifying an original as being a model worthy of copy or it may establish a new direction, break new grounds, or otherwise contribute to the progress of modernity by presenting new combinations of cultural elements" (1976, p. 81). The authenticity targeted today in heritage tourism negotiates between both these functions: first, an attempt is made to copy the original; then the copy is modified to meet the needs of the contemporary communities (Chhabra et al. 2003, p. 704). Objective authenticity (the purest version of authenticity) is popularly used as a reference point and it connotes genuineness and traditional culture from the place of its origin. The present-day authenticity pays tribute to the "original" concept and this view is reiterated by Taylor: "tourism sites, objects, images, and even people are not simply viewed as contemporaneous productions. Instead, they are positioned as signifiers of past events, epochs, or ways of life. In this way, authenticity is equated as original (2011, p. 33). Heritage tourists, often quest for an authentic experience and believe that "the authentic experience resides outside the boundary of everyday life in.... today's society. People think either the past was better or lives outside their space are better" (Chhabra et al. 2003, p. 705).

Recent literature continues to suggest that authenticity is a negotiated rather than an absolute trait of the heritage tourism spectacle. It is negotiated by a broad spectrum of stakeholders, which include the government (at local and national levels), destination marketing organizations, heritage tourism institutions, the business community, tourists, and, in the case of indigenous communities, agents from minority or silent groups. Different dimensions of the authentic can be viewed as: "commodification versus spontaneity (non-commercialization), cultural evolution versus museumfication, economic development versus cultural preservation, ethnic autonomy versus state regulation, and mass tourism development versus sustainable cultural tourism" (Wall & Xie 2005, p. 19). Tension within and between each of the dichotomies exist and perspectives of culture and heritage are created/recreated by different agencies/stakeholders who hold and/or claim a particular perspective of authenticity. What informs and directs these claims constitutes an important area of study; that is, how do numerous agents authenticate authenticity is important to know. In fact, authentication of authenticity is an emerging field of inquiry. As a scholar, I feel compelled to advance scrutiny in this field.

Equally important is the revisit of this phenomenon in the intra-pandemic times. COVID-19 is an unimaginable 'watershed' moment. Around a time, when the tourism industry was still grappling with the impacts of overtourism, the entire globe catapulted into an undertourism or a no-tourism predicament. The scale of disruption caused by COVID-19, to international tourism, is immense. According to UN-WTO, more than 95% of the countries worldwide have imposed travel restrictions and many destinations have closed their border to tourists. Some have been selective in entry permissions. Mobility and social contact form the core of the tourism and hospitality industry and the pandemic has imposed restrictions on both. It is feared that social distancing might become integrated as a cultural norm; driven by fear, warm gestures of hugs and handshakes might depart from many cultures and become a thing of the past. At the same time, a need has simultaneously emerged to resurrect/reform destinations and their offerings and strategically shape their product life cycles. This is also a wakeup call for all 'influencers and shakers' of the heritage tourism industry. As the crisis is evolving, as of end of August 2020, several countries are opening their borders cautiously and modifying their restraining measures. Heritage tourism is a crucial component of tourism with surging demand over the last several decades. Undoubtedly, many heritage institutions have had well-established agendas; but now all are at a reset level. Regardless of the turn, heritage tourism will take in the post-pandemic times; authenticity will remain its biggest moral, sustainable, commercial, and/or marketing asset. It is highly likely that it will be treated and negotiated in a more delicate manner if the new level of global consciousness remains slanted towards solidarity, virtuosity, and sustainability. Time will tell whether this awakened consciousness will sustain and produce far reaching transformations and outcomes. This book, therefore, is timely as it expands the parameters of the authenticity discourse in the context of unforeseen disasters of magnitude.

In focusing on negotiated authenticity, this book pushes the boundaries of heritage tourism research by developing a smart resilient authentication model, underpinned on ethical principles. By aiming to be smart, the purpose is to make the model both adaptable to Information and Communication Technologies (ICT) and viable for non-digital settings and experiences. Technology today is not just an interface; rather it performs a phenomenal role in shaping the supply side of heritage tourism (Hausmann & Weuster 2018) and its crucial (in fact survival) role during the pandemic times cannot be denied. At the same time, it cannot be refuted that the pre-pandemic digital era had produced a radical shift in the manner heritage offerings were developed, promoted, communicated, and sustained. More specifically, it influenced the manner in which authenticity was showcased and interpreted by different heritage tourism agencies. Emerging call for digital detoxification of heritage tourism was obtaining a strong foothold. However, the pandemic has shifted everything and crowned digitalization for facilitating connectivity; all physical touch points of socialization have been either altered or replaced. While valorizing the innovative digital boom, at this juncture, I would also like to argue that a critical discourse on digitalization should be inclusive of non-digital detox authentication perspectives of authenticity. The latter are also crucial for the development of resilient practices rooted in hands-on traditions that can open or lead to new routes of exploration to promote objective authenticity and long-term sustainability of heritage resources.

In this book, I identify and discuss the evolution and meaningfulness of negotiated authenticity in the pre-, intra-, and post-pandemic eras. Using an exploratory technique, I identify numerous case studies from across the world to examine vulnerability of a broad spectrum of heritage expressions and their resilient capabilities, particularly in the context of negotiated authenticity. Furthermore, I also scrutinize different authentication processes and offer a discourse on the manner in which authentic values are assigned to heritage objects and experiences (Wall & Xie 2005; Xie 2011). The power of authentication (both mutual and of the self) in bestowing authenticity remains a marginally explored area of study (Lugosi 2016). Furthermore, transformative solutions are needed to reform heritage tourism and bring it to the next level of moral self-authentication. It is not just the institutions and stakeholders and host communities who need to reform, but the tourists also need to deeply transform themselves beyond hedonistic and self-gratifying pursuits. They need to wake up to their inner level of consciousness and depart from hypocritical dispositions to optimize existentialist authenticity in a way that they become citizens of the world and virtuous human beings. In other words, moral selving and moralized self-authentication hold enormous potential to open new pathways toward attainment of eudaimonic well-being and social transformation, thereby fortifying resilience and sustainability. I close the book by opening this trajectory of research for future scholars.

Structure of the book

The book is divided into twelve chapters. The introduction chapter (one) offers an overview of how authenticity has evolved in heritage tourism. It presents emerging trends and discusses the manner in which they highlight the significance of authenticity in heritage tourism. It also summarizes the progress in the field of authenticity, especially in the context of its various notions, sustainability, economic value, marketing (specifically branding), and authentication.

The second chapter outlines the manner in which the notion of negotiated authenticity is developed and embraced in heritage tourism. One important contribution of this chapter is that it examines this form of authenticity from the lens of vulnerability and accords special attention to the micro and macro environment factors. As the authentication perspective gains momentum, it is important to study how heritage institutions and mediating agents authenticate heritage (Khanom, Moyle & Kennelly 2019; Wall & Xie 2005; Xie 2011). Chapter three presents the 'how' in respect to the authenticating process. The aim is also to examine the role of tangible and intangible markers and recognize mediating platforms that help shape and orchestrate the manner in which authenticity is portrayed. Digitalization continues to revolutionize heritage tourism. Chapter four shares recent trends in ICT (Information and Communication technologies) and discusses how it has or will shape authenticity of heritage tourism during COVID-19 times. It scrutinizes the role of ICT as a mediating agent in the authentication process. Furthermore, it offers insights on how ICT is employed, by heritage institutions across numerous scenarios across the globe, to promote negotiated authenticity.

Chapter five discusses negotiated authenticity and its authentication from a marketing standpoint. Undeniably, authenticity has become an important marketing tool and is being extensively used to create distinct heritage brands. Existing smart/sustainable heritage tourism frameworks are identified and examined for the manner in which they embrace negotiated authenticity. The resilience of negotiated authenticity in sustainable marketing is ascertained, especially, by taking a present-centered perspective. Next, heritage hotels and resorts have become popular cultural and historical centers for showcasing unique and especially indigenous heritage environments across the globe. Such built heritage-oriented settings significantly contribute to local and regional development in that they enrich the heritage tourism portfolio of local communities. Chapter six describes the significance of negotiated authenticity from a cultural hospitality standpoint (Chhabra 2015; Derrida 2000; Ellis 2000). Negotiated authenticity and its authentication process are discussed in the context of heritage accommodations. Also, deliberations are offered regarding the impact of the pandemic on this sector.

Homestays have emerged as an important asset of community-based tourism in rural settings across the globe to satisfy tourist quest for an authentic, novel, and personalized experience and promote sustainable consumption of cultural and rural resources. As an alternative form of accommodation that takes place in small and often remote, rural communities, its vulnerability/resilience and authentication (especially during the intra-pandemic times) are worthy of study. Chapter seven affords special attention to the authentication of homestay tourism in the Himalayas of India. Next, a review of existing literature shows that somewhat limited attention has been accorded to the authentication of national branding (Katz & Lee 1992; Pretes 2003; Smith 1991; Tang, Morrison, Lehto, Kline, & Pearce 2009). Heritage holds a significant relationship with national identity (Palmer 1999) and it is generally the most marketed aspect of a nation and an important index in the construction of national brands.

Clancy (2009) argues that nation branding by the tourism marketers influences both internal (at home) and external (abroad) audience perceptions of a nation. Authenticity of preferred narratives/icons and national identity is worthy of investigation as selective interpretations and identities continue to be debated. Chapter eight shares different meanings/images communicated/portrayed by the different stakeholders of heritage tourism and offers insights on the authentication paths pursued to brand a nation. It also offers insights on the impact of COVID-19 on how a nation's brand is perceived and the manner in which a favorable repositioning can be devised.

Museums are important repositories of heritage. They may be viewed as an early form of commodification. Several studies argue that the museums are no more the touchstones of authenticity and much debate has centered on the commodification of museums in the era of global-local nexus. Because of the changing market trends and budget cuts, museums have relaxed their traditional norms to adapt to changing environments. As pointed out by Chhabra, "museums do not exist in a vacuum. They have to reflect the current culture, and the influence of the ruling power relations" (2008, p. 428). She suggests a negotiation framework to address the fundamental problem faced by multiple and often conflicting ideologies in the contemporary era. Chapter nine extends this discourse on negotiated authenticity and its authentication by museums, especially from the pandemic standpoint. Chapter ten extends the conversation by scrutinizing the manner in which authenticity is negotiated and portrayed through selected markers in ethnic cuisines. Extant literature acknowledges that food functions as an authenticating agent for ethnic experiences (Barbas 2003; Robinson & Clifford 2012; Sims, 2009). Moreover, it is an important mechanism and provides an ideal setting to accrue cultural capital (Andersson & Mossberg 2004). Research on authenticity and authentication of cultural food offerings remains meager (Mkono 2011, 2013; Sims, 2009). According to Chhabra, Lee & Zhao, negotiated 'othered' food offerings are sought by consumers and restaurants also endeavor to negotiate their efforts to authenticate authenticity in a manner that appeals to their patrons. Patrons prefer to experience the 'other' in an objectively authentic setting but in a negotiated manner that is conducive and pleasurable to their comfort and lifestyle. This chapter examines negotiated 'otherness' of ethnic cuisines in several countries across the globe. It also offers insights on how these popularly sought heritage experiences have been impacted by the non-pharmaceutical interventions (NPIs) imposed by the pandemic.

Chapter eleven directs attention to the authentication of heritage merchandize with special focus on souvenirs. Cohen (1993, 2002) and Chhabra (2005) trace two paths of research associated with the study of souvenirs: supplier strategies (producer and vendors-retailers) and consumer preferences and behavior. The manner in which authenticity is authenticated and digitalized in the souveniring (authenticating process) process constitutes an important area of study. Insights are also offered on the way the souvenirs are taking a new meaning in the COVID-19 times.

The last chapter (twelve) 'Going Forward' offers an outline of key points raised in the book chapters and re-examines the core themes unveiled in Chapters 1–11. It presents a smart resilient negotiated authentication paradigm and offers insights, based on important lessons learned by pursuing negotiated authenticity and scrutinizing its authentication in a variety of heritage settings. It discusses the future of negotiated authenticity, from a resilient standpoint, in the context of digital/non-digital and location specific environments. It appropriates the authenticity and its authentication discourse, across different heritage expressions, in the context of the intra- and post-pandemic times. Localism and new economic order are likely to shape sustainability of heritage tourism in the future. By highlighting the sustainable prospects of mutual authentication (between guests and hosts) and moral aspects of self-authentication, the book closes with a call for expanding these trajectories of research so that they can be employed meaningfully to transform heritage tourism and relaunch it in the post-pandemic times.

References

Andersson, T. D., & Mossberg, L. (2004). The dining experience: Do restaurants satisfy customer needs? *Food Service Technology*, *4*(4), 171–177.

Barbas, S. (2003). "I'll take chop suey": Restaurants as agents of culinary and cultural change. *Journal of Popular Culture*, *36*(4), 669.

Chhabra, D. (2001). *An Analysis of Perceived Authenticity and Economic Impact of the Scottish Highland Games in North Carolina*. PhD dissertation, North Carolina State University.

Chhabra, D. (2005). Defining authenticity and its determinants: Toward an authenticity flow model. *Journal of Travel Research*, *44*(1), 64–73.

Chhabra, D. (2008). Positioning museums on an authenticity continuum. *Annals of Tourism Research*, *35*(2), 427–447.

Chhabra, D. (2012). Authenticating ethnic tourism. By P. Xie. *Tourism Management*, *33*(4), 1005–1006.

Chhabra, D. (2015). A cultural hospitality framework for heritage accommodations. *Journal of Heritage Tourism*, *20*(2), 184–190.

Chhabra, D., Healy, R., & Sills, E. (2003). Staged authenticity and heritage tourism. *Annals of Tourism Research*, *30*(3), 702–719.

Chhabra, D., Lee, W., & Zhao, S. (2013). Epitomizing the "other" in ethnic eatertainment experiences. *Leisure/Loisir*, *37*(4), 361–378.

Clancy, M. (2009). *Brand New Ireland? Tourism, development and national identity in the Irish Republic*. Ashgate Publishing, Ltd.

Cohen, E. (1993). The heterogeneization of a tourist art. *Annals of Tourism Research*, *20*(1), 138–163.

Cohen, E. (2002). Authenticity, equity and sustainability in tourism. *Journal of Sustainable Tourism*, *10*(4), 267–276.

Derrida, J., & Dufourmantelle, A. (2000). *Of hospitality*. Stanford University Press.

Ellis, J. (2000). Scheduling: the last creative act in television? *Media, Culture & Society*, *22*(1), 25–38.

Hausmann, A., & Weuster, L. (2018). Possible marketing tools for heritage tourism: the potential of implementing information and communication technology. *Journal of Heritage Tourism,* 13(3), 273–284.

Katz, H., & Lee, W. N. (1992). Oceans apart: An initial exploration of social communication differences in US and UK prime-time television advertising. *International Journal of Advertising*, *11*(1), 69–82.

Khanom, S., Moyle, B., Scott, N., & Kennelly, M. (2019). Host–guest authentication of intangible cultural heritage: a literature review and conceptual model. *Journal of Heritage Tourism,* 14(5-6), 396–408.

Kolar, T., & Zabkar, V. (2010). A consumer-based model of authenticity: An oxymoron or the foundation of cultural heritage marketing? *Tourism Management, 31*(5), 652–664.

Lee, W., & Chhabra, D. (2015). Heritage hotels and historic lodging: Perspectives on experiential marketing and sustainable culture. *Journal of Heritage Tourism*, 10(2), 103–110. https://doi.org/10.1080/1743873X.2015.1051211.

Lugosi, P. (2016). Socio-technological authentication. *Annals of Tourism Research, 58*, 100–113.

MacCannell, D. (1976). *The visitor: A new theory of the leisure class*. New York, NY: Schoken Books.

Mkono, M. (2011). The othering of food in touristic eatertainment: A netnography. *Tourist Studies, 11*(3), 253–270.

Mkono, M. (2013). Using net-based ethnography (netnography) to understand the staging and marketing of "authentic African" dining experiences to tourists at Victoria Falls. *Journal of Hospitality and Tourism Research, 37*, 184–198. doi:10.1177/1096348011425502.

Palmer, C. (1999). Tourism and the symbols of identity. *Tourism Management, 20*, 313–321.

Park, E., Choi, B. K., & Lee, T. J. (2019). The role and dimensions of authenticity in heritage tourism. Tourism Management, 74, 99–109.

Pine, B. J., & Gilmore, J. H. (2008). The eight principles of strategic authenticity. *Strategy & Leadership, 36*(3), 35–40.

Pretes, M. (2003). Tourism and nationalism. *Annals of Tourism Research, 30*(1), 125–142.

Reisinger, Y., & Steiner, C. J. (2006). Reconceptualizing object authenticity. *Annals of Tourism Research, 33*(1), 65–86. doi: 10.1016/j.annals.2005.04.003.

Robinson, R. N., & Clifford, C. (2012). Authenticity and festival foodservice experiences. *Annals of Tourism Research, 39*(2), 571–600. doi: 10.1016/j.annals.2011.06.007.

Sims, R. (2009). Food, place and authenticity: Local food and the sustainable tourism experience. *Journal of Sustainable Tourism, 17*(3), 321–336. doi: 10.1080/09669580802359293.

Smith, A. D. (1991). *National identity*. London: Penguin.

Tang, L., Morrison, A., Lehto, X., Kline, S., & Pearce, P. (2009). Effectiveness criteria for icons as tourist attractions: A comparative study between the United States and China. *Journal of Travel and Tourism Marketing, 26*, 284–302.

Taylor, J. (2001). Authenticity and sincerity in tourism. *Annals of Tourism Research, 28*, 7–26.

Wall, G., & Xie, P. (2005). Authenticating ethnic tourism: Li dancers' perspectives. *Asia Pacific Journal of Tourism Research, 10*(1), 1–21.

Wang, N. (1999). Rethinking authenticity in tourism experience. *Annals of Tourism Research, 26*(2), 349–370.

Xie, P. F. (2011). *Authenticating ethnic tourism* (Vol. 26). Channel view publications.

Acknowledgments

This book grew out of my years of work on authenticity and the authentication of heritage that began with my PhD dissertation. My interest in this topic continues to be piqued as authenticity has become a cornerstone and key sustainability criterion in heritage tourism. I am deeply engaged with scrutinizing the manner in which scholarship on this subject has progressed and evolved. From the point of time when most authenticity discourses were anchored in three dominant ideologies (objective, existentialist, and constructivist) with strategies to bridge them in a meaningful manner, to the present era where attention has shifted toward scrutinizing the processes through which objects, places, and experiences are authenticated, I remain a devoted observer and an active contributor. Recent studies have moved beyond the conceptual dissection of authenticity to examine the dynamic and multifarious process of authentication. This book is one more addition to this ocean of new knowledge and makes a dedicated effort to pen down the journey of the negotiated authenticity discourse and its authentication using multiple heritage and digital settings. I have worked to further knowledge on this topic by offering insights into how heritage can be made more resilient to survive the test of time and the disequilibrium effected by unforeseen chaotic forces. What I initially conceived as a quick and modest project turned into a lengthy and complex part of my enriching research journey. I am enormously grateful to numerous people, scholars, and heritage institutions for their assistance and support from the beginning of this work to its completion. I am also indebted to my dog, Ebony, for patiently keeping watch over me during long hours of work.

Finally, I wish to thank my father. This book is dedicated to my father, Dr. G. S. Chhabra, who is sadly no more with me but continues to inspire me through his overwhelming support and guidance, the enormous love he showered on me, and the memories we created together.

1 Introduction

> This chapter discusses the different dimensions of authenticity, its significance, and its authentication in heritage tourism. It argues that negotiated authenticity can offer a strategic pathway to support smart resilient authentication processes that are premised on ethical production and consumption of heritage. It also examines the potential of negotiated authenticity to promote resilience and sustainability and situates the discourse in the context of COVID-19.

The authenticity discourse has taken center stage in heritage tourism over the past several decades and has been the subject of various deliberations and analyses. Heritage tourism can be defined as a form of travel where travelers seek to view or experience 'built heritage, culture or modern-day arts' (Aas, Ladkin & Fletcher 2005; Frost 2006; Moscardo & Pearce 1999; Timothy 2011). It is a "phenomenon that focuses on the management of past, inheritance, and authenticity to enhance participation and satisfy consumer emotions by evoking nostalgic emotions; its underlying purpose is to stimulate monetary benefits for its various constituencies such as the museums, historic houses, festivals, heritage hotels and other stakeholders" (Chhabra 2010a, p. 5). On a positive note, heritage tourism can serve as a vehicle for conserving culture and landscape for a long time, although misuse of heritage resources (both tangible and intangible) often leads to compromise of authenticity and manipulation of the past for business goals (DeSoucey, Elliot, & Schmuz 2019; Park, Choi, & Lee 2019). Most contemporary issues in heritage tourism are associated with the following:

- forging meaningful ties between cultural heritage management (CHM) and tourism;
- viable use of heritage resources for the purpose of revenue and user-fee debate;
- visitor engagement strategies and authentic interpretation; congestion management; heritage politics (dealing with dissonance and societal amnesia);
- globalization effects (in terms of showcasing fragmented heritage);

- effective use of technology to conserve and market heritage;
- forging effective partnerships and stakeholder management; and
- managing tension between commodification and conservation of heritage and increasing demand for an authentic experience (Aas et al. 2005; Arnold 2005; Chhabra 2010a; Chhabra & Zhao 2015; Du Cros 2008, 2009; Garrod & Fyall 2001; Hede & Thyne 2007; Lowenthal 2000; McKercher & du Cros 2002; Medina 2003; Parsons & Maclaran 2009; Timothy 2011; Timur & Getz 2009).

Additionally, authenticity is itself a problematic concept with multiple meanings, appropriations, and relocation issues associated with rescuing a heritage building that can fuel a debate between its in situ preservation and re-erection procedures (Reeves, Dalton, & Pesce 2020). In the latter case, controversy emerges in the manner the meaning is contested in that it is "partially derived from its intrinsic worth and partly from how various interest groups perceive the building" (2020, p. 4). From a positive standpoint, authenticity has become a cornerstone of effective and sustainable management of heritage resources. It is being increasingly regarded as a core attribute of heritage tourism experience (Andriotis 2011; Ateljevic & Doorne 2005; Beverland 2005; Chhabra 2005, 2010b, 2012b; Chhabra, Healy, & Sills 2003, Cohen & Cohen 2012; Crang 1996; DeLyser 1999; Grayson & Martinec 2004; Halewood & Hannam 2001; Kirillova, Lehto, & Cai 2016; Rickly-Bod 2012a, 2012b; Taylor 2001) and plays a key role in attracting visitors to heritage sites (Bendix 1997; Bunce 2016; Chhabra 2010c; Kolar & Zabkar 2010; Park et al. 2019; Rickly-boyd 2013a; Timothy 2011; Xie 2011). It is regarded as a viable economic resource in that it can serve as a sustainable path to achieve smart heritage tourism development (Cavanaugh 2019; Chhabra 2015; Desoucey et al. 2019; Lugosi 2016; Pine & Gilmore 2008; Thompson & Schofield 2009; Timothy 2011; Waitt 2000; Waller & Lea 1998; Xie & Wall 2003). Smart cultural heritage tourism refers to tourism to places and sites that promote cultural heritage by embracing smart technologies, knowledge, and sustainability (such as conservation and social inclusion) (Vattano 2013). Smart heritage approach, reliant on innovative and harmonious technology, can augment the value of culture/heritage by making it more accessible, both in visual and cognitive terms. As an instance, emerging mobile technologies are offering innovative digital services that provide location- and context-specific information to tourists. Furthermore, they have the potential to offer innovative ways to regulate tourist flows, cultural heritage conservation, and beneficial social change through enhanced/meaningful relationship with the host communities. This chapter touches on key aspects of the authenticity discourse to date, especially in the context of its multiple meanings, morality, economic value, marketing, sustainability, digitalization, and the process (authentication) through which tangible and intangible heritage are endorsed. It explores the notion of resilient authentication and closes with an overview of the manner in which heritage tourism is paused and disrupted by COVID-19. Important questions are posed with regard to the foreseeable path of the authenticity discourse and its possible direction during the post-pandemic times.

Authenticity has become a driving force of tourism consumption and qualifies as a crucial benchmark for the advancement of sustainable and smart heritage tourism (Chhabra 2009b, 2010a). Its centrality in heritage tourism is undisputed (Chhabra 2010a, 2010b; Kirillova et al. 2016; Naoi 2004; Sims 2009; Timothy 2011). Some studies have shown that authenticity enriches the quality of heritage tourism (Mrđa & Carić 2019; Park et al. 2019).

Furthermore, as pointed out by Ram, Björk, and Weidenfeld (2016, p. 110), "authenticity in the context of tourism suppliers is perceived as an essential asset of firms that provide services for consumers, which are not only satisfied with low costs and high quantity, but also seek genuine experiences (Pine & Gilmore 2008)". A review of documented literature shows that authenticity has been examined immensely from both supply and demand perspectives (Chhabra 2008, 2010b; Park et al. 2019; Timothy 2011). It is a complex phenomenon because multiple views shape its discursive path. Based on antecedent viewpoints of special interest are essentialist/object (legitimate, true to the origin), constructivist (commodified for income), existentialist (optimal and euphoric), and negotiated authenticities (Chhabra et al. 2003; Chhabra, Zhao, Lee, & Okamoto 2012; Cohen 2002; DeSoucey et al. 2019; Kolar & Zabkar 2010; MacCannell 1973; 1992; Pine & Gilmore 2008; Reisinger & Steiner 2006; Wang 1999). The pure essentialist view focuses on cultural continuity. It refers to the traditional elements of culture (Taylor 2001). According to Taylor, authenticity is a kind of reproduction that holds a mirror to the original version of past. It is argued that everything authentic today is a symbol or signifier of past occurrences, eras, or ways of living (Rickly-Boyd 2012b; Salamone 1997; Timothy & Boyd 2003). In a nutshell then, the essentialist (also referred to as objective) version holds proxy to the true, original, genuine, actual, and unchanged version of heritage (de Bernardi 2019; Reeves et al. 2020; Timothy 2011). It is frozen in time and implies continuity in its most virtuous form. By the same token, several scholars also argue that essentialist authenticity is an impossible goal to accomplish (Salih 2020).

The constructivist view supports commodified forms of authenticity such as fake settings and deliberately constructed pseudo-backstages (Chhabra et al. 2003, 2012; MacCannell 1992; Medina 2003). It relates to a commodified version of culture (Trilling 1972). Because it is performed to please the markets, its true and original form gets distorted (Cohen 2002; du Cros 2009; MacCannell 1973, 1976, 1992; Silver 1993; Timothy 2011; Uriely 2005). It is argued that commodification changes the meaning of cultural markers in that they eventually become distanced from their initial worth (Chhabra et al. 2003). Next, the negotiation stance is slanted to rationalize a midway point, by retaining its originality or embracing sanitized modifications, to meet consumer demand. According to Adams (1996), this stance is a jointly constructed process between the suppliers and the consumers. Through negotiation, it is possible to safeguard and retain the core elements of object authenticity (DeSoucey et al. 2019; Jennings & Stehlik 2001; Medina 2003). Negotiation traverses between essentialist and constructivist notions; this theory argues that authenticity is not completely harmed or risked when it is modified to satisfy the market demand if it is done delicately and with caution; that is, several elements of essentialist authenticity can stay intact while adhering to the needs of the tourists market (Chhabra et al. 2003; Halewood & Hannam 2001; Waitt 2000; Wall & Xie 2005).

The existentialist view advocates optimized experience, discovery of one's true self and 'being true to oneself' (Wang 1999). It refers to exaltation and exhilaration by being just yourself and being able to express freely without inhibitions (Sloan 2007). It is purely a state of mind, hence subjective (Kim & Jamal 2007; Kirillova & Lehto 2015; Mkono 2011, 2013; Steiner & Reisinger 2006). It is key to a person's well-being and self-fulfillment (Gino, Kouchaki, & Galinsky 2015; Mknono 2020). Gino et al. argue that it is an ongoing process of personal negotiation that encompasses acknowledgment of one's "personal thoughts, emotions, needs and wants, and acting in accordance with those experiences" (Mkono 2020, p. 3).

It is centered on experiential consumption and can be delineated into two categories: intrapersonal and interpersonal. More recent work on existentialist authenticity raises scrutiny on moral grounds. Sloan (2007) writes that inauthenticity of the self is experienced when there is a disconnect between our external behavior and our real inner self. This disruption causes feelings of discomfort, dissonance, and a sense of alienation (Gino et al. 2015; Rickly-Boyd 2013b) and results in "lower moral self-regard and feelings of impurity, which trigger a desire for cleansing and prosocial, compensatory behaviors" (Mkono 2020, p. 3). In other words, when we choose to conform to our true selves, we are existentially authentic. But, if we are not able to stay true to our inner self, we are inauthentic. The pre-pandemic world has not been conducive to authentic retention because "a large portion of our existence is being lived out and shared in the digital world, where boundaries ... have become more blurry between real and fake, personal and social, private and public" (Mkono 2020, p. 3). The outcome is that a tourist can live in a state of apprehension as he or she continuously negotiates to seek harmony between the inner self and self-expression to others (Grauel 2016). Some scholars obviously argue that the existentialist state of mind is an impossible feat. In the next chapter, I will examine this position from a negotiated perspective. Furthering this probe from a technological perspective, Tribe and Mkono (2017) use the context of super-connectivity (information and communications technology [ICT]). They inquire into the manner in which super-connectivity shapes authentic experiences. The authors refer to e-lienation to examine the degree to which ICT helps or resists an alienated experience or setting.

While acknowledging the objective (essentialist), constructive and existentialist/inauthentic-self positions of authenticity, the theoplacity standpoint offers a more comprehensive and realistic approach by situating tourists in environmental and sociopolitical contexts; it recognizes the complex notion of authenticity shaped by an interactive dialogue between the visited place, trust, encounter, and the self (Belhassen, Caton, & Stewart 2008, p. 685). Several studies have used the theoplacity connotation to support negotiated positions of authenticity (Belhassen et al. 2008; Chhabra 2010b; Chhabra et al. 2012; Chhabra, Lee, Zhao, & Scott 2013; Robinson & Clifford 2012). It attaches social and cultural meanings to physical artifacts. Perusal of recent literature supports the emerging popularity and practicality of the negotiation standpoint as it seeks to reconcile between economic, cultural, and subjective worlds (Belhassen & Caton 2006, Belhassen et al. 2008; Chhabra 2008; Kirillova et al. 2016; Knudsen & Waade 2010; Lee & Chhabra 2015). In support of this view, many studies have reported that a tourist enjoys an optimized experience, feels a sense of happiness, and is in touch with himself or herself in an essentialist (original/genuine) setting. In this manner, McCabe says, "commodification can be situated within the ongoing cultural construction process" (1998, 233). Recent literature reports that embrace of authenticity (especially essentialist and negotiated versions) generates a positive value, both in economic and noneconomic terms (Chhabra 2008, 2010b; Mkono 2011; Robinson & Clifford 2012; Taylor 2001; Timothy 2011; Xie 2011; Yan 2011).

Authenticity is often touted as a key element of CHM. Authenticity from a CHM perspective is associated with portraying the past in an accurate manner (Du Cros 2001; Graburn 1989; Timothy 2011; Timothy & Boyd 2003). Against the traditional view that authenticity showcases unique cultural characteristics that have been preserved through territorial departures, the global view of commodification viewpoint implies that cultural thresholds are established to a large extent by supplier/corporate and public interests centered on monetary goals (Moscardo & Pearce 1999; Waller & Lea 1998; Wall & Xie 2005). Most fragmented perspectives of authenticity are positioned on two bipolar scales: objectivist/constructionist and existentialist/objectivist (Chhabra 2010b). Within marketing research, two research streams have evolved: authenticity as an attribute of a subject (i.e. employee's emotional authenticity; Hennig-Thurau, Groth, Paul, & Gremler 2006) and as a trait of an object (i.e. brand authenticity; Beverland, 2006).

From a marketing standpoint, extant literature recognizes that authenticity is a significant motivating driver (Frisvoll 2013; Grunewald 2002; Hughes 1995; Kolar & Zabkar 2010; Lee, Phau, Hughes, & Quintal 2015; Park et al. 2019) and has become a key selling point for heritage sites and destinations (de Bernardi 2019; DeSoucey et al. 2019; Napoli, Dickinson, Beverland, & Farrelly et al. 2014; Timothy 2011). It has become a distinct branding tool (Beverland 2006; Bryce, Curran, O'Gorman, & Taheri 2015; Chhabra 2010c; Fritz, Schoenmueller, & Bruhn 2017; Grayson & Martinec 2004; Kolar & Zabkar 2010; Morhart, Malar, Guevrement, Girardin, & Grohmann 2014). In other words, it can be used to retain consumer interests and attract potential markets through appropriate positioning and branding techniques (Bunce 2016). Having said that, it is important to identify consumer preferences for the type of authenticity sought and frame tailor-made competitive offerings (Kolar & Zabkar 2010). Several studies have confirmed that theoplacity which seeks a middle path between essentialist and existentialist types of authenticity helps explain consumer behavior in heritage tourism (Buchmann, Moore, & Fisher 2010; Chhabra 2010b; Hede & Thyne 2007). Understanding satisfaction and loyalty (such repeat visits and positive word of mouth), based on different kinds of authentic experiences, has become paramount to a destination or site's successful performance (Chhabra 2010b; Lee et al. 2015; Park et al. 2019). Using a consumption model of authenticity, Chhabra (2010b) examined relationship between perceived authenticity and tourist satisfaction at heritage sites. Her study offers strategic suggestions for proactive brand management strategies.

The dominant role of authenticity in destination image formation necessitates managers/custodians of local culture, customs, architecture, and historic landmarks to position authenticity in their branding strategies. From this standpoint, staged authenticity approach can be utilized by recreating and reenacting heritage traditions as experiencing authenticity is more about feeling one's authentic self rather than having the 'real' or 'objective' authentic experience (Moscardo & Pearce 1999). Lending a voice to this credence, Turner and Manning (1988) emphasize that the desire for authenticity is especially strong in times of change and uncertainty; it is then individuals seek safe environments that offer a sense of continuity. On the other hand, the need for authenticity is attributed to the increasing homogenization of the marketplace (Beverland & Farelly 2010). In particular, authenticity merits attention in the context of quality and differentiation in terms of market transparency and prompt flow (both viral and unidirectional) of information (Eggers, O'Dwyer, Kraus, Vallaster, & Guldenberg 2013). Informed consumers are more likely to desire consistency and authenticity in the brands they seek (Holt 2002). Brand authenticity offers a unique marketing route and has piqued the interest of numerous heritage destinations and businesses which claim to offer objectively authentic experiences.

Brand authenticity

Brand authenticity can be described as the perceived genuineness of a brand that is showcased in the form of constancy and consistency (i.e. continuity), uniqueness (i.e. originality), ability to assure (i.e. reliability), and unpretentiousness (i.e. naturalness) (Bruhn, Schoemuller, & Heinrich 2012). Numerous studies increasingly support the notion of brand authenticity as consumer quest for authenticity has become "one of the cornerstones of contemporary marketing" (Brown, Kozinets, & Sherry 2003, p. 21). According to Gilmore and Pine (2007, p. 23), "quality no longer differentiates; authenticity does". Scholarly understanding of brand authenticity is shaped by the conceptualization offered by Grayson and Martinec (2004). Building on Peirce (1991) philosophy of signs and MacCannell's (1973, 1976) distinction between 'original' (i.e. objectivist perspective) and 'staged' (i.e. constructivist perspective) authenticity, Grayson and Martinec (2004) designed a framework to examine consumer perceptions of authenticity (using indexical and iconic versions). Drawing on consumer's objective and subjective versions of authenticity, a handful studies can be identified that focus on a brand authenticity scale (Bruhn et al. 2012; Morhart et al. 2014; Napoli et al. 2014). Based on Beverland's (2006) study, Napoli et al. (2014) identify three dimensions: quality commitment, heritage, and sincerity. Morhart et al. (2014) develop a continuum based on four factors: continuity, credibility, integrity, and symbolism. Although the measurement scales depict departures (consumer's agreement of being true to themselves is only reported by Morhart and colleagues), "their operationalizations demonstrate substantial similarities, in so far as they all cover the aspects of consistency (i.e. continuity, heritage), honesty (i.e. reliability, quality commitment, credibility), and genuineness (i.e. naturalness, sincerity, integrity)" (2015, p. 5).

In summary, brand authenticity refers to the comprehended reliability of a brand's behavior, showcasing core values and patterns, so that it can be labeled as genuine. It is important to note that research on brand authenticity is still in its infancy stage. Heritage institutions and corporations require further insights on how brand authenticity can inform consumer perceptions and shape behavior. What is lacking is empirically tested suggestion, on the manner in which authenticity can be branded to help forge consumer bonds.

In this book, I argue that a brand authenticity framework for heritage tourism and its different manifestations is crucial. The purpose is to forge meaningful connections with different stakeholders of heritage tourism in a manner that supports a sustainable branding protocol. In a forthcoming chapter, I will present a brand authenticity paradigm that is underpinned on the negotiated authenticity and integrates a smart resilient component to make it sustainable (economically and culturally viable) in the long term. Support for this paradigm is captivated from the core tenets of heritage sustainability (in addition to communication mix, research, market segmentation, and environment analysis): local community involvement/benefits, economic viability, partnerships and collaboration, authenticity and conservation, interpretation, and creating mindful visitors. I argue that sustainable marketing strategies should seek to uniquely position the brand authenticity of different heritage agencies and businesses.

Furthermore, to date, very few studies have examined the economic value of authenticity in heritage tourism (Chhabra 2010a; Pine & Gilmore 2008). It can be argued that objectively or negotiated authentic forms of heritage (tangible or intangible) hold current, optional, existence, and bequest values both in economic and nonuse terms. For instance, the contemporary museum ethos can benefit from leveraging brand authenticity to support its collection and preservation goals (Bunce 2016; Chhabra 2008) and build relationships with its different publics (McLean 2012). Next, to overcome hurdles in the path of authentic branding of heritage, it is important to examine the process through which authenticity is conferred on heritage resources.

Authentication

Clearly, recent antecedent viewpoints support a middle path for authenticity. Authenticity is a negotiated rather than an absolute trait of the heritage tourism spectacle. It is negotiated by a broad spectrum of stakeholders, which include the government (at local and national levels), destination marketing organizations, heritage tourism institutions, the business community, tourists, and agents from minority or silent groups (as in the case of indigenous communities) (DeSoucey et al. 2019; Reeves et al. 2020). Different dimensions of the authentic can be viewed as: "commodification versus spontaneity (non-commercialization), cultural evolution versus museumfication, economic development versus cultural preservation, ethnic autonomy versus state regulation, and mass tourism development versus sustainable cultural tourism" (Wall and Xie 2005, p. 19). Tension within and between each of the constructs continues to exist because authenticity, particularly, in the context of heritage tourism and the self is an evolving phenomenon. What informs, guides, and shapes their viewpoints make an important area of study. That is, it is important to discern how authenticity is authenticated. Recent studies have moved beyond the conceptual dissection of authenticity and are more focused on dissecting the dynamic process of authentication.

This recent shift calls for sustainable branding of authenticity and its authentication process (Cohen & Cohen 2012; Lamont 2014; Lugosi 2016; Xie 2011). Therefore, it is important to understand different perspectives of authenticity as deciphered by consumers, heritage institutions, marketing agents such as destination marketing organizations, and the government agencies. While looking at the significance of perceived authenticity of rural tourism, Frisvoll argues "the lack of a conceptual framework through which to view and asses claims for authenticity raises the danger that we are simply reproducing popular myths countryside authenticity" (2013, p. 272).

The social side of tourist spaces (contested, negotiated, and consumed) calls for scrutiny to understand both, complexity and bias in the legitimization process and the burden of moral obligation regarding selection of true expressions (Frisvoll 2013). Cohen and Cohen (2012) stress on the need of theorizing the social routes that shape the manner in which authenticity perspectives are produced, sustained, and fortified. Xie (2011) uses Said's orientalism and Babha's third space concepts to craft an authentication approach in the context of ethnic tourism. Frisvoll (2013) defines authentication as a social process that embraces a complex spectrum of components (such as tangible and intangible products, practices, and performances) connected to conversations outside the tourism environment.

Frisvoll (2013) suggests a trialectic approach by examining the role of social representations of space and spatiality in the manner they intersect social representations (views of authenticity), materiality (as in the visual look), and practice (a tourism form such as mass, small-scale, or agricultural endeavor). In other words, he conceptualizes the process of authentication of rurality and rural space by examining the complex interplay of ideas, locality, and human practice. He studies the social process through which notions of authenticity emerge, and are gauged, and established. The 'how' and 'why' of authenticity need to be explored and in doing that, the author brings to light the "multifaceted nature of authentication meshed with materiality, social representations, political discourses, practices and performativity" (2013, p. 294). Cohen and Cohen define authentication as "a process by which something – a role, product, site, object or event – is confirmed as 'original', 'genuine', 'real', or 'trustworthy'" (2012, p. 1296). The authors delineate between cool authentication and hot authentication and argue that these two types of authentication are related to different kinds of authentic experiences. Cool authentication is a kind of endorsement by which authenticity of an object is confirmed to be original and real, rather than fake. Hot authentication, on the other hand, "is an immanent, reiterative, informal performative process of creating, preserving, and reinforcing an object, site's or event's authenticity" (2012, p. 1300).

Zhu contends that "authentication has become a governance strategy to legitimize inclusion and exclusion and to allocate economic, moral and aesthetics values" (2014, p. 12). Aligned to this, Xie (2011) examines authentication in the context of power relations and authority exercised by the role players. Multiple insights are needed to critically deconstruct and construct these areas of exploration to enrich understanding of authenticity and its authentication in heritage tourism. Clearly, the key notions of authenticity continue to be shaped by different players, who, in turn, are impacted by the broader environment beyond their control. Several studies stress on the need to explore authentication (social processes) to inspect the manner in which authentic values are imbued on heritage objects and experiences (Chhabra et al. 2013; Xie 2011; Wall & Xie 2005). Also, the role of power and authority in the authentication of authenticity remains a marginally explored area of study (Lugosi 2016).

The entire interplay of politics in authentication is an outcome of "controversy and contestation" (Cohen & Cohen 2012, p. 1306). Selected role players or stakeholders/agencies hold or propagate the authority to authenticate and power dynamics differ in hot and cool authentication in that it is more specific in cool instances (as power to authenticate is vested in a few persons or agencies) and fragmented in the hot authentication process. According to Cohen and Cohen, "the political questioning regarding cool authentication is how power is obtained and how it is exercised or contested", for instance, "cool authenticating procedures deployed by experts and institutions are sometimes contested by other experts, leading to controversy regarding the authenticity of given objects, sites or events" (2012, p. 1307).

This book seeks to broaden the discursive parameters of authenticity in all its totality of expressions and identify power mechanisms that shape the manner in which it is produced, marketed, and consumed. This is an attempt to share contemporary views on how the contemporary notions of authenticity are derived, interpreted, applied, processed, and legitimized in local and global contexts. Given the scholarly progress in this field, it is surprising that academic dialogue on authentication of heritage is still meager. In summary, the take away points from this chapter are that negotiated positions are noted in both heritage tourism supply and demand environments. It is posited that negotiated authenticity can offer a strategic pathway to support smart resilient authentication processes that are premised on ethical production and consumption of heritage. By striving to be smart, institutions show willingness to embrace Internet Communication Technology (ICT). A critical discourse of digitalization and digital authentication of authenticity remains an unexplored area of study. Technology today is not just an interface; rather it has played a phenomenal role in shaping the supply side of heritage tourism (Chhabra 2015). The contemporary digital era has produced a radical shift in the manner heritage offerings are developed, promoted, and communicated. More specifically, it has influenced the manner in which authenticity is showcased and interpreted by heritage tourism suppliers. A critical discourse of digitalization and digital authentication of authenticity remains an unexplored area of study.

In closing, situating the authenticity and authentication debate in the context of the current pandemic (COVID-19) is a complex task. The coronavirus pandemic has interrupted the entire socioeconomic structures across the globe and has deeply impacted the travel and tourism industry (Gossling, Hall, & Scott 2020; Higgins-Desbiolles 2020). COVID-19 has presented an unparallel catastrophe to the tourism sector, especially because of travel embargos, border shutdowns, and quarantine regulations. The United Nations World Tourism Organization (UNWTO) is coordinating closely with the World Health Organization (WHO), its Member States, and tourism industry. Many World Heritage Sites across the globe have closed their doors, halting intangible cultural heritage performances, which has produced socioeconomic outcomes for host communities. The cultural heritage tourism sector holds tremendous potential to contribute in terms of recovery and reformation efforts. Recent studies have brought several suggestions to the table such as promotion of degrowth strategies, strengthening local supply chains, and promotion of local products (Gills 2020; Gossling, Hall, & Scott 2020; Higgins-Desbiolles 2020; Ranasinghe et al. 2020). It is hoped that the new social order will unfold "new forms of collective human consciousness; a new type of global social covenant; new forms of appropriate technology; and new forms of appropriate lifestyle" (Gills 2020, p. 579). The novel transcending routes should be centered on the long-term well-being and resilience of tourists, hosts, and the complete heritage tourism system.

In the following chapters, I will deliberate on questions such as: What it means to offer or have an authentic experience today? What shape will authenticity take when people start traveling tomorrow or in some countries where travel has commenced? It is likely that authenticity will take new meanings as reformative cultural travel pathways are planned in the foreseeable future.

References

Aas, C., Ladkin, S., & Fletcher, J. (2005). Stakeholder collaboration and heritage management. *Annals of Tourism Research*, *32*(1), 28–48.

Adams, V. (1996). *Tigers of the show and other virtual Sherpas: An ethnography of Himalayan encounters*. Princeton, NJ: Princeton University Press.

Andriotis, K. (2011). Genres of heritage authenticity: Denotations from a pilgrimage landscape. *Annals of Tourism Research*, *38*(4), 1613–1633.

Arnold, D. (2005). Virtual tourism – a niche in cultural heritage. In Novelli, M. (Ed.), *Niche tourism: Contemporary issues, trends and cases*. Burlington, MA: Elsevier.

Ateljevic, I., & Doorne, S. (2005). Dialectics of authentication: Performing 'exotic otherness' in a backpacker enclave of Dali, China. *Journal of Tourism and Cultural Change*, *3*(1), 1–17.

Belhassen, Y., & Caton, K. (2006). Authenticity matters. *Annals of Tourism Research*, *33*(3), 853–856.

Belhassen, Y., Caton, K., & Stewart, W. P. (2008). The search for authenticity in the pilgrim experience. *Annals of Tourism Research*, *35*(3), 668–689.

Bendix, R. (1997). *Search of authenticity: The formation of folklore studies*. Madison, WI: University of Wisconsin Press.

Beverland, M. (2005). Crafting brand authenticity: The case of luxury wines. *Journal of Management Studies*, *42*(5), 1003–1029.

Beverland, M. (2006). The 'real thing': Brand authenticity in the luxury wine trade. *Journal of Business Research*, *59*(2), 251–258.

Beverland, M. B., & Farrelly, F. J. (2010). The quest for authenticity in consumption: Consumers' purposive choice of authentic cues to shape experienced outcomes. *Journal of consumer research*, *36*(5), 838–856.

Brown, S., Kozinets, R., & Sherry, J. (2003). Teaching old new tricks and the revival of brand meaning. *Journal of Marketing*, *67*(3), 19–33.

Bruhn, M., Schoemuller, V., & Heinrich, D. (2012). Brand authenticity: Toward a deeper understanding of its conceptualization and measurement. *Advances of Consumer Research*, *40*, 567–576.

Bryce, D., Curran, R., O'Gorman, K., & Taheri, B. (2015). Visitors' engagement and authenticity: Japanese heritage consumption. *Tourism Management*, *46*, 571–581.

Buchmann, A., Moore, K., & Fisher, D. (2010). Experiencing film tourism: Authenticity & fellowship. *Annals of Tourism Research*, *37*(1), 229–248.

Bunce, L. (2016). Appreciation of authenticity promotes curiosity: Implications for object-based learning in museums. *Journal of Museum Education*, *41*(3), 230–239.

Cavanaugh, J. R. (2019). Labelling authenticity, or, how I almost got arrested in an Italian supermarket. *Semiotic Review*, (5).

Chhabra, D. (2005). Understanding authenticity and its determinants. *Journal of Travel Research*, *44*(1), 64–73.

Chhabra, D. (2008). Positioning museums on an authenticity continuum. *Annals of Tourism Research*, *35*(2), 427–447.

Chhabra, D. (2009a). Sustainable marketing of unique museums. *Asian Journal of Tourism and Hospitality Research*, *3*(2), 78–83.

Chhabra, D. (2009b). Proposing a sustainable marketing framework for heritage tourism. *Journal of Sustainable Tourism*, *13*(3), 303–326.

Chhabra, D. (2010a). *Sustainable marketing of cultural and heritage tourism*. London: Routledge.

Chhabra, D. (2010b). Back to the past: Generation Y's perceptions of authenticity. *Journal of Sustainable Tourism*, *18*(6), 793–809.

Chhabra, D. (2010c). Branding authenticity. *Tourism Analysis*, *15*(6), 735–740.

Chhabra, D. (2012a). A presented-centered dissonant heritage management model. *Annals of Tourism Research*, *39*(3), 1701–1705.

Chhabra, D. (2012b). Authenticity of the objectively authentic. *Annals of Tourism Research*, *39*(1), 499–502.

Chhabra, D. (2015). A cultural hospitality framework for heritage accommodations. *Journal of Heritage Tourism*, *10*(2), 184–190.

Chhabra, D., & Zhao, S. (2015). Present-centered dialogue with heritage representations *Annals of Tourism Research*, *55*, 94–109.

Chhabra, D., Healy, R., & Sills, E. (2003). Staged authenticity and heritage tourism. *Annals of Tourism Research*, *30*(3), 702–719.

Chhabra, D., Lee, W., Zhao, S., & Scott, K. (2013). Marketing of ethnic food experiences: Authentication analysis of Indian cuisine abroad. *Journal of Heritage Tourism*, *8*(2/3), 145–157.

Chhabra, D., Zhao, S., Lee, W., & Okamoto, N. (2012). Negotiated self-authenticated experience and homeland travel loyalty: Implications for relationship marketing. *Anatolia: International Journal of Hospitality and Tourism Research*, *23*(3), 429–436.

Cohen, E. (2002). Authenticity, equity and sustainability in tourism. *Journal of Sustainable Tourism*, *10*(4), 267–276.

Cohen, E., & Cohen, S. A. (2012). Authentication: Hot and cool. *Annals of Tourism Research*, *39*, 1295–1314.

Crang, M. (1996). Magic kingdom or a quixotic quest for authenticity? *Annals of Tourism Research*, *23*(2), 415–431.

de Bernardi, C. (2019). Authenticity as a compromise: A critical discourse analysis of Sámi tourism websites. *Journal of Heritage Tourism*, *14*(3), 249–262.

DeLyser, D. (1999). Authenticity on the ground: Engaging the past in a California ghost town. *Annals of the Association of American Geographers*, *89*(4), 602–632.

DeSoucey, M., Elliott, M. A., & Schmutz, V. (2019). Rationalized authenticity and the transnational spread of intangible cultural heritage. *Poetics*, *75*. doi: 10.1016/j.poetic.2018.11.001

Du Cros, H. (2001). A new model to assist in planning for sustainable cultural heritage tourism. *International Journal of Tourism Research*, *3*(2), 165–170.

Du Cros, H. (2008). Too much of a good thing? Visitor congestion management issues for popular world heritage tourist attractions. *Journal of Heritage Tourism*, *2*(3), 225–238.

Du Cros, H. (2009). Emerging issues for cultural tourism in Macau. *Journal of Current Chinese Affairs*, *38*(1).

Eggers, F., O'Dwyer, M., Kraus, S., Vallaster, C., & Guldenberg, S. (2013). The impact of brand authenticity and SME growth: A CEO perspective. *Journal of World Business*, *48*(3), 340–348.

Frisvoll, S. (2013). Conceptualizing authentication of ruralness. *Annals of Tourism Research*, *43*, 272–296.

Fritz, K., Schoenmueller, V., & Bruhn, M. (2017). Authenticity in branding-exploring antecedents and consequences of brand authenticity. *European Journal of Marketing*, *51*(2), 324–348.

Frost, W. (2006). Braveheart-ed Ned Kelly: Historic films, heritage tourism and destination image. *Tourism Management*, *27*(2), 247–254.

Garrod, B., & Fyall, A. (2001). Heritage tourism: A question of definition. *Annals of tourism research*, *28*(4), 1049–1052.

Gilmore, J., & Pine, B. (2007). *Authenticity: What consumers really want?* Boston, MA: Harvard Business School Press.

Gills, B. (2020). Deep restoration: From the great implosion to the great awakening. *Globalizations*, *17*(4), 577–579.

Gino, F., Kouchaki, M., & Galinsky, A. D. (2015). The moral virtue of authenticity: How inauthenticity produces feelings of immorality and impurity. *Psychological Science*, *26*(7), 983–996.

Gössling, S., Scott, D., & Hall, C. M. (2020). Pandemics, tourism and global change: A rapid assessment of COVID-19. *Journal of Sustainable Tourism*, *29*(1), 1–20.

Graburn, N. (1989). Tourism: The sacred journey. In V. Smith (Ed.), *Hosts and guests* (2nd ed., pp. 21–52). Philadelphia, PA: University of Pennsylvania Press.

Grauel, J. (2016). Being authentic or being responsible? Food consumption, morality and the presentation of self. *Journal of Consumer Culture*, *16*(3), 852–869.

Grayson, K., & Martinec, R. (2004). Consumer perceptions of iconicity and indexicality and their influence on assessments of authentic market offerings. *Journal of Consumer Research*, *31*(2), 296–312.

Grunewald, R. (2002). Tourism and cultural revival. *Annals of Tourism Research*, *29*(4), 1004–1021.

Halewood, C., & Hannam, K. (2001). Viking heritage tourism: Authenticity and commodification. *Annals of Tourism Research*, *28*(3), 565–580.

Hede, A., & Thyne, M., (2007). *Authenticity and branding for literary heritage attractions*. ANZMA Papers: University of Otago.

Hennig-Thurau, T., Groth, M., Paul, M., & Gremler, D. (2006). Are all smiles created equal? How emotional contagion and emotional labor affect service relationships. *Journal of Marketing*, *70*(3), 58–73.

Higgins-Desbiolles, F. (2020). Socialising tourism for social and ecological justice after COVID-19. *Tourism Geographies*, *22*(3), 610–623.

Holt, D. (2002). Why do brands cause trouble? A dialectical theory of consumer culture and branding. *Journal of Consumer Research*, *29*(1), 70–90.

Hughes, G. (1995). Authenticity in tourism. *Annals of Tourism Research*, *22*(4), 781–803.

Jennings, G., & Stehlik, D. (2001). Mediated authenticity: The perspective of farm tourism providers. Paper presented at the 32nd Annual Conference of the Travel and Tourism Research Association, Fort Myers, Florida.

Kim, H., & Jamal, T. (2007). Touristic quest for existential authenticity. *Annals of Tourism Research*, *34*(1), 181–201.

Kirillova, K., & Lehto, X. (2015). An existential conceptualization of the vacation cycle. *Annals of Tourism Research*, *55*, 110–123.

Kirillova, K., Lehto, X., & Cai, L. (2016, ahead of print). Existential authenticity and anxiety as outcomes: The tourist in the experience economy. *International Journal of Tourism Research*. Wiley Online Library. doi: 10.1002/jtr.2080

Knudsen, B. T., & Waade, A. M. (2010). Performative authenticity in tourism and spatial experience: Rethinking the relations between travel, place and emotion. In B. T. Knudsen & A. M. Waade (Eds.), *Re-investing authenticity: Tourism, place and emotions* (pp. 1–21). Bristol: Channel View Publications.

Kolar, T., & Zabkar, V. (2010). A consumer-based model of authenticity: An oxymoron or the foundation of cultural heritage marketing? *Tourism Management*, *31*(5), 652–664.

Lamont, M. (2014). Authentication in sport tourism. *Annals of Tourism Research*, *45*, 1017.

Lee, S., Phau, I., Hughes, M., & Quintal, V. (2015). Heritage tourism in Singapore Chinatown: A perceived value approach to authenticity and satisfaction. *Journal of Travel and Tourism Marketing*, *33*(7), 981–998.

Lee, W., & Chhabra, D. (2015). Heritage hotels and historic lodging: Perspectives on experiential marketing and sustainable culture. *Journal of Heritage Tourism*, *10*(2), 103–110.

Lowenthal, D. (2000). *The past is a foreign country*. Cambridge: Cambridge University Press.

Lugosi, P. (2016). Socio-technological authentication. *Annals of Tourism Research*, *58*, 100–113.

MacCannell, D. (1973). Staged authenticity: Arrangements of social space in tourist settings. *American Journal of Sociology*, *79*, 589–603.

MacCannell, D. (1976). *The tourist: A new theory of the leisure class*. New York, NY: Schocken Books.

McCabe, S. (1998). Contesting home: Tourism, memory, and identity in sackville, New Brunswick. *The Canadian Geographer*, *42*(3), 231–245.

MacCannell, D. (1992). Cannibalism today. In D. MacCannell (Ed.), *Empty meeting grounds*: The tourist papers (pp. 17–73). London: Routledge.

McKercher, B., & du Cros, H. (2002). *Cultural tourism: The partnership between tourism and cultural heritage management*. Binghamton, NY: Haworth Press.

McLean, F. (2012). *Marketing the museum*. London: Routledge.

Medina, L. (2003). Commoditizing culture: Tourism and Maya identity. *Annals of Tourism Research*, *30*, 353–368.

Mkono, M. (2011). The othering of food in touristic eatertainment: A netnography. *Tourist Studies*, *11*(3), 253–270. doi: 10.1177/1468797611431502.

Mkono, M. (2013). Using net-based ethnography (netnography) to understand the staging and marketing of "authentic African" dining experiences to tourists at Victoria falls. *Journal of Hospitality and Tourism Research*, *37*, 184–198. doi:10.1177/1096348011425502.

Mkono, M. (2020). Eco-hypocrisy and inauthenticity: Criticisms and confessions of the eco-conscious tourist/traveller. *Annals of Tourism Research*, *84*, 102967.

Morhart, F., Malar, L., Guevrement, A., Girardin, F., & Grohmann, B. (2014). Brand authenticity: An integrative framework and measurement scale. *Journal of Consumer Psychology*, *25*(2), 200–218.

Moscardo, G., & Pearce, P. (1999). Understanding ethnic tourists. *Annals of Tourism Research*, *26*(2), 416–434.

Mrđa, A., & Carić, H. (2019). Models of heritage tourism sustainable planning. In *Cultural urban heritage* (pp. 165–180). Cham: Springer.

Naoi, T. (2004). Visitors' evaluation of a historical district: The roles of authenticity and manipulation. *Tourism and Hospitality Research*, *5*(1), 45–63.

Napoli, J., Dickinson, S., Beverland, M., & Farrelly, F. (2014). Measuring consumer-based brand authenticity. *Journal of Business Research*, *67*, 1090–1098.

Park, E., Choi, B. K., & Lee, T. J. (2019). The role and dimensions of authenticity in heritage tourism. *Tourism Management*, *74*, 99–109.

Parsons, E., & Maclaran, P. (2009). *Contemporary issues in marketing and consumer behaviour*. London: Routledge.

Peirce, C. (1991). "On the Nature of Signs". In Hoopes J. (Ed.), *Peirce on Signs: Writings on Semiotic by Charles Sanders Peirce* (pp. 141-143). University of North Carolina Press. Retrieved February 21, 2020, from www.jstor.org/stable/10.5149/9781469616810_hoopes.12.

Pine, B. J., & Gilmore, J. H. (2008). The eight principles of strategic authenticity. *Strategy & Leadership*, *36*(3), 35–40.

Ram, Y., Björk, P., & Weidenfeld, A. (2016). Authenticity and place attachment of major visitor attractions. *Tourism Management*, *52*, 110–122.

Ranasinghe, R., Karunarathna, C., & Pradeepamali, J. (2020). After Corona (COVID-19) impacts on global poverty and recovery of tourism-based service economies: An appraisal. Available at SSRN 3591259.

Reeves, C. D., Dalton, R. C., & Pesce, G. (2020). Context and knowledge for functional buildings from the industrial revolution using heritage railway signal boxes as an exemplar. *The Historic Environment: Policy & Practice*, 1–26.

Reisinger, Y., & Steiner, C. J. (2006). Reconceptualizing object authenticity. *Annals of Tourism Research*, *33*(1), 65–86. doi: 10.1016/j.annals.2005.04.003.

Rickly-Boyd, J. (2013a). Aleination: Authenticity's forgotten cousin. *Annals of Tourism Research*, *40*(1), 412–415.

Rickly-Boyd, J. (2013b). Existential authenticity: Place matters. *Tourism Geographies*, *15*(4), 680–686.

Rickly-Boyd, J. M. (2012a). Authenticity & aura: A Benjaminian approach to tourism. *Annals of Tourism Research*, *39*(1), 269–289.

Rickly-Boyd, J. M. (2012b). 'Through the magic of authentic reproduction': Tourists' perceptions of authenticity in a pioneer village. *Journal of Heritage Tourism*, *7*(2), 127–144.

Robinson, R. N., & Clifford, C. (2012). Authenticity and festival foodservice experiences. *Annals of Tourism Research*, *39*(2), 571–600. doi: 10.1016/j.annals.2011.06.007.

Salamone, F. (1997). Authenticity in tourism: The San Angel inns. *Annals of Tourism Research*, *24*(2), 305–321.

Salih, S. (2020). Cinematic authenticity-effects and medieval art: A paradox. In *Medieval film*. Manchester: Manchester University Press.

Silver, I. (1993). Marketing authenticity in third world countries. *Annals of Tourism Research*, *20*, 302–318.

Sims, R. (2009). Food, place and authenticity: Local food and the sustainable tourism experience. *Journal of Sustainable Tourism*, *17*(3), 321–336.

Sloan, M. (2007). The "real self" and inauthenticity: The importance of self-concept anchorage for emotional experiences in the workplace. *Social Psychology Quarterly*, *70*(3), 305–318.

Steiner, Y., & Reisinger, C. (2006). Understanding existential authenticity. *Annals of Tourism Research*, *33*(2), 299–318.

Taylor, J. (2001). Authenticity and sincerity in tourism. *Annals of Tourism Research*, *28*(1), 7–26.

Thompson, K., & Schofield, P. (2009). Segmenting and profiling visitors to the Ulaanbaatar Naadam festival by motivation. *Event Management*, *13*(1), 1–15.

Timothy, D. (2011). *Cultural and heritage tourism*. Bristol: Channel View Publications.

Timothy, D., & Boyd, S. (2003). *Heritage tourism*. Harlow, England: Pearson.

Timur, S., & Getz, D. (2009). Sustainable tourism development: How do destination stakeholders perceive sustainable urban tourism? *Sustainable Development*, *17*(4), 220–232.

Tribe, J., & Mkono, M. (2017). Not such smart tourism? The concept of e-lienation. *Annals of Tourism Research*, *66*, 105–115.

Trilling, L. (1972). *Sincerity and authenticity*. London: Oxford University Press.

Turner, C., & Manning, P. (1988). Placing authenticity – On being a tourist: A reply to Pearce and Moscardo. *The Australian and New Zealand Journal of Sociology*, *24*(1), 136–139.

Uriely, N. (2005). The tourist experience: Conceptual development. *Annals of Tourism Research*, *32*(1), 199–216.

Vattano, S. (2013). European and Italian experience of smart cities: A model for the smart planning of city built. *TECHNE-Journal of Technology for Architecture and Environment*, 110–116. https://doi.org/10.13128/Techne-12809

Waitt, G. (2000). Consuming heritage: Perceived historical authenticity. *Annals of Tourism Research*, *27*(4), 835–862.

Wall, G., & Xie, P. F. (2005). Authenticating ethnic tourism: Li dancers' perspectives. *Asia Pacific Journal of Tourism Research*, *10*(1), 1–21.

Waller, J., & Lea, S. (1998). Seeking the real Spain: Authenticity in motivation. *Annals of Tourism Research*, *25*(4), 110–129.

Wang, N. (1999). Rethinking authenticity in tourism experience. *Annals of Tourism Research*, *26*, 349–370.

Xie, F., & Wall, G. (2003). Visitors' perceptions of authenticity at cultural attractions in Hainan, China. *International Journal of Tourism Research*, *4*, 353–366.

Xie, P. (2011). *Authenticating ethnic tourism*. Bristol: Channel View Publications.

Yan, L. (2011). Authenticating ethnic tourism. *Journal of Heritage Tourism*, *6*(3), 267–268.

2 Negotiated authenticity and vulnerability

This chapter draws attention to the limitations and impossibility of offering the purest version of authenticity. It also examines how different forms of authenticity continue to be re-ordered and institutionalized and how negotiated authenticity is subject to cross examination by human agency. Using the notion of governmentality, this chapter scrutinizes the manner in which authenticity is compromised or commodified or kept close to the original version. In doing so, it seeks to look afresh at the paradox that shapes the traversed layers and parameters of authenticity.

Recent literature, to a vast extent, outlines that authenticity is perceived, experienced, and negotiated differently by different people. In his seminal work on authenticity, Jones (2010) questions why authenticity is obsessively sought and what could be the possible purpose of this quest. He proclaims that several dynamics are at work and some might be guided by cultural or biological traits while others are based on a network of relationships people form with objects, places, and/or other people. Authenticity has been extensively viewed, both from a materialistic standpoint and as a culturally constructed phenomenon. The latter position claims that authenticity is always in a state of flux and evolves as culture evolves. It is a mixed value bestowed by markets or suppliers or other (public and private) stakeholders who have a vested interest. This implies that authenticity is not fixed but fluid in nature. All recognized or bestowed authenticity is an outcome of a tradeoff or a mediation. It is now commonplace to note that the negotiated stance has become the magnet of discussion in contemporary heritage tourism literature. Simultaneously, a need has emerged to explore/identify numerous processes of authentication that legitimize authenticity. These processes are shaped by various institutions, stakeholders, and markets and are subject to power dynamics. Furthermore, in the quest to remain close to the original, vulnerability positions of negotiated authenticity constitute an important field of inquiry.

This chapter draws attention to the limitations and impossibility of offering the purest version of authenticity. It also examines how different forms of authenticity continue to be reordered and institutionalized and how negotiated authenticity is subject to cross examination by human agency. Culture and heritage evolve with time and cultural norms are vulnerable to power dynamics and changing political, social, and economic environments. It can be argued that authenticity is cumulatively authorized through the process of governmentality. Foucault (1970) wrote that governmentality refers to shared power knowledge and collective privileges. Using the notion of governmentality, this chapter scrutinizes the manner in which authenticity is compromised or commodified or kept close to the original. In doing so, it seeks to look afresh at the paradox that shapes the traversed parameters of authenticity. Three strands guide this chapter:

- Dialogical nature of authenticity
- Relativization of negotiated authenticity
- Heritage vulnerability and negotiated authenticity

Vast literature now confirms that authenticity of culture and heritage in tourism is a dialectical process. In other words, culture and heritage are presented through several masks/layers of authenticity, some of which converge while others are in opposition with each other. To understand how authenticity is negotiated and how this negotiation grows or changes over a definite or an indefinite period of time, it is impertinent to identify its definers, producers, influencers, carriers, and receivers. For instance, some authors have fetishized on different carriers/custodians of cultures and probed into the interdependencies and power struggles associated with the selection and showcasing of authenticity. There is a need to comprehend how authenticity is molded by cultural industries/agents while keeping it relevant within the consumer culture. Contrary to the claims by numerous scholars that the concept of existentialist authenticity was introduced by Wang in the late twentieth century and that negotiated authenticity became the touch point of scholarly dialogue in the twenty-first century, few authors have linked the negotiated dispositioning and emergence/support for the existentialist school of thought to earlier times; it was during these times, that the traditional views of culture and heritage began to be interrogated and this questioning created a shift-distancing of 'self' from its cultural layers. Origins of existentialist authenticity can be found in "old philosophical traditions of fully expressing one's individual self in a social world, asserting one's social choices, living in harmony with one's own sense of self, and being attuned with one's individual experiences, rather than seeing .... through agency structures" (Mkono 2020, p. 2). Vast body of work exists on this topic that spans several eras. For instance, deliberation of these concepts can be noted in the seminal work of authors like Freud, Focault, Bhabha, Lowenthal, Trilling, and Hollinshead.

Later, Ashworth and Urry enshrined the negotiated standpoint. This chapter offers further ponderance of their work in the context of the authenticity debate.

The dialogical nature of authenticity

Authenticity is dialectical, partly, because of the evolving nature of culture (Freud & Freud 1992). Trilling (1965) in his book 'beyond culture' interprets

Freud's insights on how culture is negotiated in the context of reality and pleasure principles. The pleasure principle is the psychoanalytic notion, drawn from the pleasure drive of the id, where people pursue gratification and elude sorrow in order to fulfill their biological and psychological needs. On the contrary, the reality principle refers to our cognitive ability to gauge the outside reality and act appropriately (Freud & Freud 1992). The theme of contradiction between the two themes is centered around the notions of love and power: one's own "cultural commitment and power .... the biological self and the culture" (Trilling 1965, pp. 96, 101) While discussing Freud's interpretation of culture, Trilling points out that the:

> cultural change happens in the mind, which then is transformed and shaped by the manner in which culture develops. Freud opined that culture plays a key role in formation of 'self' and that the self transcends the world of culture bestowed on us. He makes us reflect on how people are conditioned by culture and helps us distance it from the self. He argues that society plays a key role in grounding people in the cultural environment and shapes their beliefs and attitudes. However, our responses can vary. If culture and society are amiable, they dominate and influence otherwise, the self can oppose the cultural norms. The manner of our opposition can be shaped by our culture. Freud offers a discourse on how our self has a tendency to connect to culture and how it should relate to it (1965, p. 108).

Foucault closely scrutinized the manner in which dominant values influenced and constrained the human mind. His notion of governmentality is underpinned on power dynamics and institutional policies. Interpreting Foucault's (1970) stance on the *governmentality of things*, Hollinshead writes:

1 It points to the need of understanding the manner in which human beings present themselves as subjects and treat others as objects. Discursive knowledge guides human experience and social existence of human beings and objects;
2 Power works within and across the institution enwrapped in knowledge, it imprisons both the ones who dominate and the subdued;
3 Within institutions, the manner in which things are contemplated and interpreted is not static. Tunbridge and Ashworth (1996) raise interesting questions associated with 'whose heritage is it and .... who interprets it and for what purpose';
4 As relationships between knowledge and power shape or generate new forms, they produce living systems that define cultural, political, or administrative exclusion;
5 Foucault argues that when people relate or identify themselves with the contemporary era, they get simultaneously confined.... to the subjective and practical nature of their own meaning making. In tourism, there can be difficult psychic states in which populations become entangled.

> As an instance, the manner in which a particular indigenous community handles the agonistics of its current development (spurred by tourism) and its welcome appreciation of the new accessibility of consumer goods while simultaneously, mourns its loss of appreciation of sacred/spiritual objects. Tourism studies have not yet thrown up anyone seemingly keen or able to comprehensively inspect the self-disciplinary profiles of the encounter with the other in tourism: tourism studies still need its full time inspectors of the monologues of cultural representation (1999, pp. 15–18).

In the context of negotiated authenticity, it can be argued that power dynamics sway the dialogue between celebration and remorse. That is, change is celebrated while lamenting about disintegration of the fixed. But it is important to note that claims over objective authenticity representations are embedded in social and political relations between external and external human agencies. Therefore, power dynamics play a key role in defining authenticity and determining how it should be authenticated. This power is a byproduct of metaphysics and collective human practice. Of relevance here is the concept of panopticon which pins attention to the knowledge-power dichotomy (Foucault 1970). As an example, constant surveillance of objects, happens inside the museums, by cultural custodians and curators. One of the earliest commentators who connected panopticon to tourism was Urry:

> by introducing Foucault's concept of surveillance into tourism, Urry does not so much enable or encourage 'host' or 'foreign populations' to be studied, but he enables and encourages tourism administrators, tourism managers and tourism 'professionals' to be studied almost as if they were foreign population acting with the domain of tourism. In other words, Urry suggests investigations into 'mental territories' and psychic selves. Representations of savage populations are made to seem like that through mundane surveillances of ourselves acting as orthodox tourists and serving as mainstream tourism planners. Such representations are conceivably the modalities of power and of the exclusionary forces of held knowledges within the everyday and conventional disciplinary business of tourism management/development/research: we all utter (1999, p. 15).

Undeniably, the postmodern era has stimulated the discourse on ourselves and made us aware of our subjective dispositions, but it is important to note that these notions were also nurtured in the premodern era. The capacity of self-intervention toward a more sustainable behavior is heightened today and interrogation of dominant discourses has transformed the debate on authenticity. Coexistence between the cultural environment and one's true or optimal or exhilarated self opens up space for reconciliation. Negotiation, in the context of culture and heritage, can be framed successfully if a person is adaptive or at his/her best and in harmony with the surrounding environment (Freud 2012). Numerous factors have shaped the manner in which authenticity of culture and heritage and the self continues to be negotiated. In regard to the interpretation of culture, two perspectives warranty attention. The first view points to the complex spectrum of activities, which encompass the practice of arts and relevant intellectual discourses (Hollinshead 1999). This notion molds the way we differentiate popular culture from high culture (endorsed by the government institutions). The second notion is more inclusive and is swayed by the macro- and micro-environment factors. For instance, culture can be sculpted by technology, mannerisms, customs, religious beliefs, and reference organizations. In the context of developing countries with rich pre-era cultural and heritage resources, historical discourses chaperone existing policies whereas in developed countries "social organization and economy is often found to be distanced from the concept of culture" (Hollinshead 1999, p. XI). These polarized examples support the tents of governmentality and illustrate that cultures are predisposed because of a wide range of factors such as history, the role of different human agencies, sociocultural traits of relevant power brokers and political/economic circumstances.

From a heritage tourism perspective, culture and heritage are commodified to charm the market. Trilling (1965) argues that demand for myths elevates the position of objects/experiences and shapes the nature and content of performance and its showcasing. In other words, because the real story or object might not be pleasing to the eye, it is softened to make it charming and personable. Several studies have argued that objects and their stories are based on dominant discourses and require appropriate matching. Hollinshead posits that, "objects play the role of actors and knowledge illuminates them. It is all about intelligibility and duality of doing versus showing, demonstrating vs performing and presenting versus representing" (1999, p. 4).

Reflection on the authenticity phenomena also unfolds several complexities associated with change and continuity. Ongoing negotiation interrogates actuality, redefines it, and adds a new perspective to its presentation. According to Trilling, "while it looks old, heritage is actually something new. It is a mode of cultural production in the present that has recourse to the past" (1965, p. 108). From another standpoint, the redefined version of heritage can breathe life into dying economies and unfamous sites. Several examples are noted to illustrate this process:

> Curatorial interventions may attempt to rectify the errors of history and make the heritage production a better place than the historical actuality it represents. Theatrical performances of heritage in developing countries exemplify the strategic use of interface to convey messages of modernity that stand in contrast with the heritage on display. To hide the interface is to foster the illusion of no mediation, to produce 'tourist

> realism' which is itself a highly mediated effect. The very term folklore signifies a special relationship to what is designated whether that relationship is marked by burlesque, nostalgia, irony or dismay. Errors become safe as they are negotiated and acquire archaic and exotic values (Trilling 1955, p. 8).

Interpretation implies that negotiated initiatives adapt actuality in a manner that preferred messages can be communicated to targeted audience. The process draws on perspectives and reactions of the audience and uses a creative/artistic approach to modify. Curatorial mediation is evidence of compromised initiatives to correct historical fallacies and produce/present heritage in an attractive manner, although the reality may be horrifying in nature. Clearly authenticity is a negotiated phenomenon because human mind (Freud & Freud 1992) is constantly seeking mediation within the vibrancy of time and culture. This negotiation does not necessarily compromise the genuineness of culture; it can serve as a sustainable medium to sensitively and meaningfully embrace change.

As a case in point, the need for urban heritage areas to evolve and experience sociocultural transformation is acknowledged in the NARA Document on Authenticity, ICOMOS (Pendlebury, Short, & While 2009). The vibrant and dynamic nature of authenticity is also addressed in the Vienna Memorandum (2005) and can be evidenced from the principles guiding the need to assimilate a historic perspective in contemporary architecture. García-Esparza writes, "given that, the term 'value' was recognized by the Nara Document to determine authenticity (ICOMOS 1994) in relation to original and subsequent characteristics of the cultural heritage, several authors have linked this term with the social construction of a given time and place. This means that value involves understanding the nature of the valued object – referred as static authenticity – and the nature of the value expressed for an object can be referred to as dynamic authenticity" (García-Esparza 2016, p. 24).

Relativization of negotiated authenticity

Undoubtedly, the floating dynamics relate to the negotiated notion of authenticity, in that the discourse is about perception, performance, feelings, and social habits, about worthiness of time and location, about bodies altering in space instead of blending with space (García-Esparza 2016). Furthermore, the object of attention is "directly affected in its materiality and composition by decision-making processes and it is the output of cumulative sociocultural construction of specific cultures. The value resides in how the object reflects the circumstances rather than in the importance of the element itself" (García-Esparza 2016, p. 24). Static, proxy term for objective authenticity, forms the core of a conservation ethic. Negotiated (dynamic) authenticity is the outcome of complex and broader/extrinsic social-constructivist processes. According to Jones (2010), the fluidity refers to the evolving authentic self and aligned with this view, García-Esparza writes that "the human nature contains elements of constancy and change in relation to biological nature, and perception and behavior, which are culturally linked and therefore, changeable, innate and constant" (2016, p. 29). In this context, it can be deduced that authenticity of culture/heritage is not a fixed objective state because the aim is to either to reach close to actuality or modify to connect with or entertain the audience. Different cultural and heritage tourism markets desire different things. One segment of the cultural and heritage tourism audience may pursue harmony, both between the self and the actual, as it quests for authenticity whereas another market might value hedonic experiences.

Next, it is important to scrutinize which preferred social/cultural practices are sustained when a specific form of authenticity is selected, that is: how authenticity is ordered/authorized, owned, purified, and/or commodified. Several cultural and heritage institutions continue to pursue versions of originality. However, it cannot be denied that the original has also evolved and is shaped by social-constructivist factors. That is, the 'actuality' (or a fixed point) changes when it is presented, performed, or interpreted (Jones 2010). This standpoint valorizes the dialectic path of negotiated authenticity, especially in the context of cultural conservation. So, one may pose a question: what is the ideal desired form of static or objective authenticity that can be protected from deterioration and is that state even possible? Is it preservation in its current form or restoration to its original state? Tangible core of authenticity and the transformation of some of its materialist aspects can lend credence to the constructivist form. In this context, Jones writes that:

> The materialist stance treats authenticity as a dimension of nature with real and immutable characteristics that can be identified and measured. On the other hand, there is the constructivist position, popular amongst academics and cultural critics, who see authenticity as a product of 'culture', or 'to be precise, the many different cultures through which it is constructed. Wrapping different layers of authenticity, one can question the popularity of some replica over others. How and why some become more powerful loci of authenticity than others, one might ask. Furthermore, to what extent is their authenticity a product of their physical state and material substance? (2010, pp. 182–183).

The term rationalized authenticity was introduced by DeSoucey et al. (2019) to refer to procedures that aim to water down incongruity between diversity celebrations of certain cultural forms and the use of formally regulated processes to authorize and defend their universal bearing. Speaking of 'universalization of particularization', DeSoucey et al. argue that such enshrining enriches the status and cultural assets of unique places and promotes collective accountability at the national stage. By the same token, universal conventions aim to mold these local or national heritage treasures within the collective universal agenda by illuminating their significance as history to be shared with the humanity as large. The UNESCO's formal world heritage site (WHS) nomination process "facilitates claims of irreplaceable heritage from nominators outside the core and enshrines their practices as universally meaningful by creating a type of authenticity that aims to balance originality and conformity. In our view, attention to instances of rationalized authenticity can shed light on how and why some elements of intangible heritage spread and potentially influence global tastes" (DeSoucey et al. 2019, p. 3). Rationalized authenticity, therefore, can take multiple configurations.

This process of rationalizing authenticity under the UNESCO Intangible Cultural Heritage (ICH) Convention illustrates how authenticity is negotiated by particularizing select aspects of folk traditions (ICH) and mounting them on a global pedestal as an exemplar and for admiration. They are "promoted as protecting threads of ancestral ways of life", the entire initiative is to imbue cultural practices with past localized values that resonate with the present and hint at preferred futures that are meaningful and of value to humanity at large. The authors remark that:

> "the operation of cultural diversity at the global level rests on an interesting dualism – we are compelled to recognize and celebrate special differences, but at the same time, diversity is something we all have in common. Heritage can offer symbolic and ...... monetary benefits to places that .... illuminate cultural histories and practices that make them distinct, supporting narratives of imagined communities as phenomena of collective and commemorative belonging .... these ongoing practices are open to negotiation, change, and innovation... each item is a reified container of social and organizational relationships, making them usable to the groups that nominated them and secondarily to the public" (DeSoucey et al. 2019, p. 9).

In this context, rationalized or negotiated authenticity serves as a process through which select elements of ICH are universalized and conflict between the traditional and the innovative mechanisms are resolved. Clearly, authenticity is relativized and lives of all objects go through dynamic social, political, and economic processes. This implies that there is no 'absolute reality' (Kim et al. 2019). The World Heritage Convention's stance of authenticity, in the Nara document (calling for unbiased and comprehensive insights on cultural diversity and cultural heritage, associated with conservation, to appraise the value and authenticity of cultural property), echoes a Eurocentric (western ideology) approach and propels attention toward power dynamics and the authoritarian position of the nominee.

Moving forward, taking a demand-side perspective, Bryce, Murdy, and Alexander (2017) state that authenticity is a process that continues to be negotiated between a tourist site and the tourist. This vibrancy often takes place at/around the sites of authenticity and is mostly referred as an interactive or co-created process (Cohen & Cohen 2012). Indeed, most literature points to numerous middle/rationalized grounds between the main strands of authenticity: existentialist, essentialist, and constructivist (materialistic) as resonance is sought between old and new meanings bestowed by cultural entrepreneurs and other mediating institutions such as UNESCO. In other words, there is no fixed point or position of authenticity. The point remains in a state of flux and several middle points might be emerge simultaneously depending on the nature of the mediating institutions (local, national, or global) and the role of cultural entrepreneurs.

Authenticity thus becomes a rationalized process that is conceptually premised "on the transnational spread, legitimation, and formalized adaptations of these practices while they are still rooted in specific (though not static) senses of authenticity linked to their country of origin" (DeSoucey et al. 2019, p. 10)

From an existentialist standpoint, "the study of any sacred site is the study of objective authenticity from the believer's point of view" (Belhassen, Caton, & Stewart 2008, p. 686) and authenticity is a cultural construction based on self-judgment. Thus, an individual's perception of authenticity occurs through a filter of personal thoughts, anticipation, and actual experience. Wang makes a crisp statement that objects may "appear authentic not because they are inherently authentic, but because they are constructed as such in terms of points of view, beliefs, perspectives, or powers" (1999, p. 351). Negotiation of actuality is molded by several factors, some of which might operate simultaneously. A notable example in this regard is the "conflict between the preservationist ethos of a WHS and attempts by local authorities to extract economic benefit or at least secure appropriate economic and social development" (Pendlebury et al. 2009, p. 350). The accelerated growth of WHS is often propelled by economic and political goals and ironically, these often deny the inclusion of local voices. Dissonant views and tension between the local and the universal can be noted across the world, as the global accolade and maintenance measures are set by international bodies. But the sites are managed by local people and their values often conflict with the universal agenda (DeSoucey et al. 2019). Aligned with this view, Pendlebury explains:

> The pressure to present heritage locations in ways deemed suitable by the tourism industry, to commodify them for tourist consumption, raises tensions with management objectives centered around notions of cultural authenticity. The range of factors to consider and mediate are many in the case of urban WHS: the extent of WHS boundaries and buffer zones; large numbers of landowners and stakeholders; and conflicts between tourism development and planning for the local population. The concept remains sketchy at its best. Thus, whilst WHS may

> sit at the pinnacle of international conservation regime, there is significant room for maneuver as this comes to ground in particular places. This multi-scaled negotiation is evident in each of the cases we describe, where, to different degrees, pro-development local and national administration is involved in an on-going debate with UNESCO and ICOMOS over questions about the significance of the site and the most appropriate form of WHS management (2009, p. 357).

In another notable discourse on the complex nature of authenticity, Jones ponders on the problematic dichotomy between materialist and constructivist viewpoints and illuminates compelling authenticity issues. Clash, between authentic conservation and commodification practices, is not new. However, a somewhat righteous middle point between the two can be touched that has enough valorized portion of objective authenticity. However, such integral positionings follow a complicated route. García-Esparza writes that:

> Each scenario, county, region, landscape and cultural asset needs to embrace a compromising path toward preservation and sustainability but the question is how. Today, in international terms, it is widely accepted that the vernacular involves rural streetscapes, field or landscape patterns, traditional uses, memories, senses, economy and culture. Nonetheless, there is a vibrant discourse about heritage resiliency, critical heritage … … and heritage in transition or … … by appropriation. All these contrast with 'the statutory', traditional lobbies, canonical texts based on the scientist materialism stemming from conservation theories, and quite a few authors have analyzed the issue from the visual experience of the 19th and 20th century canonical perceptions … often referred to as the leisure consumption of heritage. (2016, p. 22)

When people experience and negotiate authenticity through objects, it is the network of relationships between people, places, and things that appear to be at the core, not the objects themselves. According to Jones (2010):

> people use connections between people, places and objects which fortifies the captivating power of authenticity and helps to traverse their place in the universe. Most persistent reference point for describing generic authenticity is the objective standpoint. It is a popular reference point and refers to authenticity as a facet of nature with real and absolute traits that can be identified and appraised. Not least of these is the historic environment, where authenticity haunts the practices of preservation, curation, management, and presentation enacted on monuments, buildings, places and artefacts. Until recently, approaches to authenticity in heritage management and conservation have been characterized by an overwhelmingly materialist perspective. Authenticity is

seen as an objective and measurable attribute inherent in the material fabric, form and function of artefacts and monuments, and a positivist set of research methods and criteria have evolved to test their genuineness. Furthermore, these approaches guide the core approach of heritage conservation and management.

On the other hand, recent scholarship outside the realm of heritage management and conservation has focused on exploring the complex and dynamic nature of authenticity and its cultural connotations. The core idea propelled is that authenticity is a culturally constructed phenomenon and its meaning is shaped by 'who is observing the object and in what context' (Jones 2010, p. 181). In the later context then, authenticity evolves and Jones writes: "it is all about disappointment, debunks and fakeness, people work with objects and places to develop and strengthen social networks and relationships in a meaningful way. We need a means to understand the powerful, almost primordial, discourses that often draw on material qualities of stone and soil, roots and nourishment, and which ultimately seem involved in working out genuine or truthful relationships between objects, people and places. We need to ask why people find ideas of authenticity so compelling and what social practices and relationships these ideas sustain. We also need to return to the materiality of objects, sites and places – an aspect that has been rather neglected by constructivist critiques and indeed by much of the recent research focusing on the experience of heritage" (2010, pp. 182–183).

Furthermore, societies in the contemporary era have become multicultural and super diverse. With the advent of social media, multiple 'micro-hegemonies' now reside in our individual selves. That is, we have developed multiple personalities and selves. The authors refer to a person's life span as an individual life project which is home to "complex micro hegemonies within which subjects situate their practices and behavior" (2011, p. 3). Blommaert and Varis (2011) offer unique insights into the heuristics of authenticity, from the standpoint of superdiversity. The authors scrutinize 'enoughness' of a content to produce enough material to experience authenticity. Identities can be described in pluralistic terms. A four-point framework is presented to offer insights into these identity processes: (1) discursive orientations toward a symbolic set of features or traits; (2) encountering of configured features; (3) varying degrees of absorption; and (4) familiarity when reregistrations are dialogically positioned (or aligned).

Therefore, identity procedures embody "stratified distinctions between 'experts' and 'novices', 'teachers', and 'learners' and 'degrees' of authenticity; contestations and conflicts centered on enoughness" (Blommaert & Varis 2011, p. 4).

As a case in point, 'Britishness' can appear in a different form every time it is showcased, expressed, or experienced prompting more contesting appropriations. As set criteria are embraced, some people might understand them better than others while others might encounter ambiguity when they seek harmony with the selected attributes. The 'enregistering' term is used here instead of registering because it refers to a particular manner of configuration that keeps changing and people repeat the process of registration instead of acquiring it and learning about it becomes a one-time process. Enoughness, on the other hand, delimits the parameters and serves as a benchmark for accepting an 'identity' type that is drawn from objective criteria (real and true to the original). According to Blommaert and Varis, "this is a slippery terrain as 'enough' is manifestly a judgment, often a compromise and rarely a 'black and white' and well-defined set of criteria for a judgment call of sorts" (2011, p. 5). The authors further note that the heritage language can suffice as an act of authentic identity. Greetings and other concise communicative rituals, indigenous songs or dances that make up for the absent culture can produce 'enough' authenticity. Demonstrating 'enoughness', and employing it as a tool to showcase/communicate indigenous authenticity, in the face of changing sociocultural and technological environments, supports a middle point where an acceptable trade-off path (between change and continuity) can be mapped. Further justification is that:

> In enregistering such features, certain rules need to be observed for the process to be successful – to be recognized by others as what was intended. These are the rules that 'newcomers', 'beginners', 'wannabes' need to observe and mobilize in their own identity work in order to 'pass' as authentic to someone. This is where Internet, for all the freedom and opportunity, is seen as offering for creative identity-play;
>
> How authenticity is manufactured by blending a variety of features, some of the defining ones – are sufficient to produce the particular targeted authentic identity;
>
> In different niches of our social and cultural lives, we arrange features in such a way that they enable others to identify us as 'authentic', 'real', members of social groups even if this authenticity comes with a lower rank as 'apprentice' within a particular field.
>
> Throughout all this, we see that 'cultures' as things that are perpetually subject to learning practices. One is never a 'full' member of any cultural system, because the configurations of features are perpetually changing and one's fluency of yesterday need not guarantee fluency tomorrow;
>
> Power is operating at a variety of scale levels in a polycentric sociocultural environment in which all of us, all the time, are required to satisfy the rules of recognizability. It enables an anti-essentialist framework that, however, does not lapse into a rhetoric of fragmentation and contradiction, but attempts to provide a realistic account of identity practices. Such practices, one will observe in a variety of domains, revolve around a complex and unpredictable notion of authenticity, which in turn, rests on the notion of 'enoughness.' The concise framework sketched here can

serve as a heuristic for engaging this enormous and rapidly changing domain of authenticity. (Blommaert & Varis 2011, pp. 6–12)

Negotiated authenticity, then, is a web of compelling floating notions woven around fluidity, enoughness, enregistering, constancy, continuity, and rationalizations. Enoughness can be referred to a fixed point where negotiation is temporarily rationalized and positioned, only to move when equilibrium shifts in the face of changing circumstances. In other words, this fixed point shifts to a new position as culture and environments evolve but this resting point is transitional. In this manner, negotiated authenticity becomes a never-ending story, striving to retain enough objective attributes while rationalizing the modifications to the past or continuity expressions to resonate with the present goals/needs of mediating agencies and the multifaceted audiences.

Parallel to this stream of thought, Fagence (2019) offers an insightful view of the manner in which heritage tourism engages with the never-ending process of telling stories. The author illustrates how past and present are negotiated in the narratives of historical sites and objects.

Semiotics are often used to decipher fixed points in narratives, which use both tangible and visual markers, to tell the story. Stories are operationalized and in the process of authenticating a never-ending story, fixed points are mapped to make the narrative as objectively authentic as possible. It cannot be denied that power discourses shape the manner in which a story is told and claimed to be authentic; a time travel experience is offered by strategically establishing a conduit between the past and present. Fagence argues that "a story told through tourism will seldom be permanently fixed, and this circumstance has implications for some of the ideological interpretations of 'authenticity', and although the fundamental shape of the story might not be imperiled, the quality of the experience presented through tourism might be" (2019, pp. 1–2).

Moving forward, because objective authenticity of heritage is commodified to adapt to market demand, it remains vulnerable in that it runs the risk of abandoning its core value. Studies on the vulnerability phenomenon, in the context of heritage and/or authenticity, are rare although heritage and authenticity have been scrutinized from sustainability or sustainable livelihood viewpoints, as evidenced in the case of ecotourism or rural tourism or indigenous tourism (Calgaro, Dominey-Howes, & Lloyd 2014). Vulnerability of both objectively authentic heritage and its negotiated version warranties scholarly attention as this research direction can help identify weak or sensitive touch points that might damage/destroy the very existence of the actual or original culture or heritage.

Vulnerability

Vulnerability can be described as "sensitivity of systems to various forms of stress and agency of human actors" (Cochrane 2010, p. 1). Turner, Kasperson, Matson, McCarthy, Corell et al. refer to vulnerability as "the degree to which a system, subsystem, or system component is likely to experience harm due to exposure to a hazard, either a perturbation or stress/stressor" (2003, p. 8074). Calgaro et al. (2014) further postulate that vulnerability is specific to a place or a system and is heavily graded, evolving, and varied. For instance, stress, reaction, and ability to adapt to change over an extended period of time can be used as metrics to demonstrate the extent to which a heritage source is vulnerable (Adger 2000). Calgero et al. highlight three interconnected dimensions that shape adaptation patterns: exposure, sensitivity, and adaptive capacity. An individual or a place or an object's vulnerability can be gauged based on the ability to anticipate, withstand, and recover from shocks over time; Facilitating factors can be level of control and access to sociopolitical, economic, and environmental resources and authority of mediating systems. Limited access and control over resources are often dictated by: (1) the conflicting initiatives and preferences of multiple agencies/platers; (2) the strength and effectiveness of different layers of governance systems and social connections (that offer access to some while limit privileges to others); and (3) historically grounded power mechanisms and cultural patterns/principles reinforcing ideologies (revealing a biased display of prerogatives) that shape selected pathways of development (Calgaro et al. 2014).

Furthermore, there are three categories of sensitivity dimensions that reinforce the notion of vulnerability: "physical, social or institutional exposure to stress; sensitivity to stress or perturbation including the ability to anticipate and cope, depending on political, social and institutional characteristics; and the ability of a system to recover, based on existing structures and to adapt to these structures to withstand future perturbations more successfully" (Cochrane 2010, p. 3). Drawing from documented literature, Calgaro et al. (2014) identifies 12 factors that can accelerate vulnerability of a tourism destination:

- Access to resources
- Highly seasonal
- Livelihood dependency
- Ecologically sensitive and hazard-prone
- Place specific
- Destination remoteness or inaccessibility
- Institutional flexibility
- Travel motivations and consumer choices
- Reliance on external marketing
- Limited disaster preparedness
- Image sensitivity to risk

Vulnerability to a large extent can be used to explain/justify the different negotiated authenticity positions, especially when the aim is to conserve/protect fragile or threatened heritage. For instance, in the case of museums, artifacts are vulnerable to both conservation resources/management (supply) techniques and market dynamics. Seasonal nature of demand can deplete a museum of its alternate source of income, in the face of dwindling public funds and make it vulnerable to compromising initiatives that can be damage the core value (or genuineness/originality) of exhibits. In the context of homestay tourism, the host can become vulnerable to changing demand, inaccessibility, support of public sector, and other important stakeholders. Macro-environment factors such as weather, politics, and melting (or assimilated) cultures can induce commodification strategies that are distanced from family and cultural values. Lack of training and the nature of the location can also make negotiated authenticity of a heritage resource vulnerable to commercialization. That is, service quality can be compromised and, in case of remote location, tourists interested in sustainable and serious leisure pursuits might not find their way to the location. All these factors can impede initiatives that aim at reviving dying handicrafts or restrict aid to economically sustain traditional livelihoods in remote regions.

Another case in point is ethnic cuisines. Negotiated authenticity of food heritage and ethnic cuisines are vulnerable to external factors. Their coping mechanisms and their adaptive capacity are reliant on marketing strategies, human resource management, and competition from other eateries including fast food. Vulnerability is also impelled if coordination between key stakeholders and public sector support is remiss or weak, for instance, if linkages are not strong between food, transportation, and accommodation sectors. Important heritage tourism decisions should be guided by the depth and integration of these connections. Significant partnerships with host community are crucial, especially if local heritage is showcased as a tourism commodity. Examples, in this category, specifically include heritage hotels, homestay tourism, and ethnic tourism.

Finally, discourse on vulnerability will be incomplete without mentioning resilience. According to Walker et al. (2004), resilience is the ability of a system to absorb shock and restore itself from a chaotic state. It refers to a system's ability to reconfigure itself to provide continuity to an authentic purpose, and/or identity. In other words, resilience is the ability to survive in the face of turbulence and disequilibrium. It is capability to counter vulnerable situations and higher the resilience, weaker the risk. Purest form of authenticity is not possible because culture continues to evolve to adapt to the changing environment. Therefore, the negotiated authenticity position holds optimal potential to make culture and heritage assets resilient and sustainable because it seeks opportunities, to strategically harmonize continuity with change, in an ongoing manner.

Closing comments

How one perceives the point of fixity or the point of *'original' or the 'objective' authenticity* can depend on how it is expressed and the position it holds in the mind of a person (visitor or any other type of heritage audience). It is also shaped by an institution's political or economic agenda and can be evidenced in the way directional messages are developed and transmitted. Latham argues that "the point of fixity is floating; it moves depending on a set of circumstances surrounding the person-document transaction. It could be logically placed (e.g. a clearly identified change in context) or arbitrarily placed (e.g. critics and experts interpret the start of a movement)" (2016, p. 2). As discussed in this chapter, fixity is the unchanging state of a heritage resource and authenticity is inherently intertwined with the notion of moving fixity, especially in its negotiated form. Latham further explains that:

> Floating fixity depends on agreed upon fixed points (socially, extrinsic) and perceived fixed points (adtrinsic). For example, in museums, fixity is still heavily integrated (and unquestioned) into the entire enterprise. To illustrate, a museum may accept an object into their collection, and (most often) the next step is to designate what it is (e.g. using guides such as Chenall's Nomenclature or some other taxonomy), where it comes from, when it was made, and so on. Many of these characteristics are intrinsic, others are extrinsic and adtrinsic (provenance) but they are not sorted out this way. Once this object is accessioned into the collection then, it has been given an identity. This identity is often singular (e.g. A wagon wheel from 1864 made of wood and metal) or at least, is layered but from a single core. Then, the object is used in an exhibition. It is here that its identity can change based on what is needed for the exhibit. Once the object enters the public sphere in an exhibition or program, it can represent many meanings and may have multiple personalities/identities depending on how it has been situated and who might be viewing it. What is interesting here is that objects are often treated as monosemic (having only a single meaning) from any one party's standpoint, but in reality these add up to the object being polysemic (having multiple meanings). Clearly there is no single fixed point for an object (museum document). Who decides which is the correct one? While it may come into the museum as a certain type of thing and subsequently assigned an identity by the registrar or collection manager, once the object enters the public sphere, the point of fixity could change; its "origin" can vary, depending on the viewer's intention. Yet, museums often function as if an object=one meaning. If we could sort out the different kinds of information in each document (or potential information based on document transactions), richer, more complex identities can arise from museum collections. In museums, assumptions about authenticity as a fixed and defined point need to be continually questioned and deeply considered. (2016, pp. 5–6)

In other words, floating fixity or negotiated authenticity requires ongoing scrutiny. Strategic resilience strategies are required to withstand vulnerable nature of negotiated authenticity, particularly, when authenticity is undergoing a sanitization process, to keep the core or original essence of culture/heritage intact, while adapting to change. Although dismissal of materialistic or objective authenticity is supported be several authors, recent research suggests that specific tangible attributes of historical objects actually contribute to the manner in which people experience and rationalize authenticity (Jones & Yarrow 2013). Particularly, "aging, patina and material decay are significant elements in the experience of authenticity", as these create a subtle sense of 'historicity or pastness', for instance, "the feelings transmitted by these visual signs of age participate in the construction of a relationship to time and form a significant element in the replication of archaeological places where the ambiance", created through technological tools, holds potential to offer deep time travel experiences (Duval, Smith, Gauchan, Mayer, & Malgat 2020, p. 143). In the contemporary era, ongoing resilience move toward digital mediums for heritage collection and dissemination can be evidenced. According to Duval et al., "by combining sensory experiences and digital reality, hybrid devices can enrich in situ visits through the use of augmented reality, consisting of new interactive interpretive techniques, which shape the manner in which cultural engagement for the public is offered" (2019, p. 142). In the contemporary era, virtual and augmented reality can generate and strengthen resilience through community bonds and sustainable cultural production designs and showcasing. Innovative projects like CURIOS (Beel, Wallace, Webster, Nguyen, Tait et al. 2017) also aid in building resilient communities:

> CURIOS uses an interdisciplinary approach to help community heritage volunteers maintain a digital presence that is sustainable overtime – how resilient behavior is enacted through cultural activity, which in turn has led to the desire to develop digital collections. It is an application that represents a form of action research which through its development use digital technologies as a mechanism that can enable rural communities to be more resilient through enhancing existing practices. The need for using digital technology is one in which the communities themselves identify as the next step in their on-going practices and represents a way in which to push their collections beyond their locality. It is shown how the historical societies represent groups of volunteers attempting to articulate their own narrative history that is largely driven by their collective sense of place; and in the process of doing so, have become further reaching in terms of their initiative to build resilience, embrace change and forge community bonds through historical production (Beel et al. 2017, p. 460).

In light of the above, resilience, in the context of heritage tourism, can be framed around the capacity of human agency to diminish cultural vulnerability to macro- and microshocks.

Resilient strategies can foster the ability to support negotiated commodification underpinned on sustainability principles. The overarching aim is to ensure that objectively authentic cultural resources have a viable economic value and this illuminates a paradox. Cavanaugh writes that capitalism and authoritarian regimes and protocol such as intellectual property structure the fiscal value of authenticity. This affords access to "manuality, time (i.e., tradition), and place (i.e., locality) to larger-scale producers whose heritage making strategies may be automated and industrial in nature, but who can afford to disseminate, profit from, and protect these signs in their own packaging and labeling .... However, the paradox here is that that those who produce 'authentic' heritage foods may be the least-well positioned to profit from them" (Cavanaugh 2019, p. 1).

The voice of local cultural bearers is often dimmed because of internal/external politics during the selection of preferred authentic embodiment of their heritage. As suggested by Beel et al. (2017), the need is for 'entrepreneurial heritage agencies' to garner the support of both heritage institutions (such as historical societies) and local communities at the grassroots level (by integrating marginalized voices) to assist in building culturally rich repertoires of local knowledge. Negotiation is as much about excavating, protecting, and passing on the past to future generations as it is about unfreezing and developing strategically/rationally 'authenticized' cultural/heritage resources for present use. Situating this discourse in today's reality shaped by COVID-19, ongoing negotiation of different heritage agencies is evident in the manner digital platforms are being leveraged for training and to situate history and heritage in the context of present times. The 'Bridge to Crafts Careers' program is a notable example of how heritage and conservation agencies are leveraging technology to promote preservation of heritage:

> In New York City, the training of young stone masons from underserved communities at the 'Bridge to Crafts Careers' program is one of the most important activities. In response to today's new reality, new sessions have been switched to virtual online classes for their required safety training. This technical instruction is essential for these young adults to obtain their certification and find a job in the next few months. In 2015, World Monuments Fund launched the Bridge to Crafts Careers, an initiative established to provide training in the preservation trades. The program offers underrepresented young adults in the New York City area hands-on technical training with the opportunity for placement in a stable career. (WHF 2020)

Several initiatives can be noted of how past is reframed and made meaningful by offering creative exercises to inspire and engage audience as they stay confined to their homes. One such example is the 'Virtual Slow Art at Home: Guided Meditation with Phoenix Art Museum':

> Hosted by a Phoenix Art Museum educator, Remix It: Virtual Summer Workshops for Teens encourages art lovers aged 13 to 18 to create their own mixed-media artworks inspired by an object from the Museum's collection. Each session of this workshop series will explore a different approach to the creative process. Sessions will be presented through Zoom, a video conference platform. (Phoenix Art Museum 2020)

Although most heritage and cultural attractions have become physically unreachable today, the current environment is conducive to recognize the potential and promise of more localized heritage expressions from the grass roots level by digitally engaging a broad spectrum of audience. Innovatively negotiated transformations and showcasing should happen with utmost sensitivity so that the core of past and heritage is not compromised. In the next chapter, I turn to the authentication process and offer a critical interrogation of its prevailing discourses. I also scrutinize the authenticating process to offer deeper insights into the manner negotiated authenticity is authenticated and its prospects and challenges in the pandemic era.

References

Adger, W. (2000). Social and ecological resistance: Are they related? *Progress Human Geography*, *24*(3), 347–364.

Andriotis, K. (2011). Genres of heritage authenticity: Denotations from a pilgrimage landscape. *Annals of Tourism Research*, *38*(4), 1613–1633.

Beel, D. E., Wallace, C. D., Webster, G., Nguyen, H., Tait, E., Macleod, M., & Mellish, C. (2017). Cultural resilience: The production of rural community heritage, digital archives and the role of volunteers. *Journal of Rural Studies*, *54*, 459–468.

Belhassen, Y., Caton, K., & Stewart, W. P. (2008). The search for authenticity in the pilgrim experience. *Annals of Tourism Research*, *35*(3), 668–689.

Blommaert, J., & Varis, P. (2011). Enough is enough: The heuristics of authenticity in superdiversity. *Tilburg Papers in Culture Studies*, *2*, 1–13.

Bryce, D., Murdy, S., & Alexander, M. (2017). Diaspora, authenticity and the imagined past. *Annals of Tourism Research*, *66*, 49–60.

Calgaro, E., Dominey-Howes, D., & Lloyd, K. (2014). Application of the destination sustainability framework to explore the drivers of vulnerability and resilience in Thailand following the 2004 Indian Ocean tsunami. *Journal of Sustainable Tourism*, *22*(3), 361–383.

Cavanaugh, J. R. (2019). Labelling authenticity, or, how I almost got arrested in an Italian supermarket. *Semiotic Review*, (5).

Cochrane, J. (2010). The sphere of tourism resilience. *Tourism Recreation Research*, *35*(2), 173–185.

Cohen, E., & Cohen, S. A. (2012). Authentication: Hot and cool. *Annals of Tourism Research*, *39*(3), 1295–1314.
DeSoucey, M., Elliott, M. A., & Schmutz, V. (2019). Rationalized authenticity and the transnational spread of intangible cultural heritage. *Poetics*, *75*, 101332.
Duval, M., Smith, B., Gauchon, C., Mayer, L., & Malgat, C. (2020). "I have visited the Chauvet Cave": The heritage experience of a rock art replica. *International Journal of Heritage Studies*, *26*(2), 142–162.
Fagence, M. (2019). Using geographical and semiotic means to establish fixed points of a never-ending story: Searching for parameters of authenticity in a case study of Australian history. *Journal of Heritage Tourism*, 1–13.
Foucault, M. (1970). *The order of things: An archaeology of the human sciences*. New York, NY: Vintage Books.
Freud, S. (2012). *The basic writings of Sigmund Freud*. Modern library.
Freud, S., & Freud, E. L. (1992). *Letters of Sigmund Freud*. Courier Corporation.
García-Esparza, J. (2016). Re-thinking the validity of the past. Deconstructing what authenticity and integrity mean to the fruition of cultural heritage. *VITRUVIO-International Journal of Architectural Technology and Sustainability*, *1*(1), 21–34.
Hollinshead, K. (1999). Tourism as public culture: Horne's ideological commentary on the legerdemain of tourism. *International Journal of Tourism Research*, *1*(4), 267–292.
ICOMOS (1994). *The Nara document on authenticity*. Paris: ICOMOS.
Jones, S. (2010). Negotiating authentic objects and authentic selves: Beyond the deconstruction of authenticity. *Journal of Material Culture*, *15*(2), 181–203.
Jones, S., & Yarrow, T. (2013). Crafting authenticity: An ethnography of conservation practice. *Journal of Material Culture*, *18*(1), 3–26.
Kim, S., Whitford, M., & Arcodia, C. (2019). Development of intangible cultural heritage as a sustainable tourism resource: The intangible cultural heritage practitioners' perspectives. *Journal of Heritage Tourism*, *14*(5–6), 422–435.
Mkono, M. (2020). Eco-hypocrisy and inauthenticity: Criticisms and confessions of the eco-conscious tourist/traveller. *Annals of Tourism Research*, *84*, 102967.
Pendlebury, J., Short, M., & While, A. (2009). Urban world heritage sites and the problem of authenticity. *Cities*, *26*(6), 349–358.
Phoenix Art Museum (2020). Slow Art at Home. Retrieved on May 2020 from: https://phxart.org/may-2020-virtual-events-hosted-by-phoenix-art-museum/
Trilling, L. (1955). Freud and the crisis of our culture. Retrieved from http://psycnet.apa.org
Trilling, L. (1965). *Beyond culture: Essays on literature and learning* (p. xii). New York, NY: Viking Press.
Tunbridge, J. E., & Ashworth, G. J. (1996). *Dissonant heritage: The management of the past as a resource in conflict*. John Wiley & Sons.
Vienna Memorandum (2005). *World heritage*. UNESCO. Retrieved from: https://whc.unesco.org/en/documents/5965
Walker, B., Holling, C. S., Carpenter, S. R., & Kinzig, A. (2004). Resilience, adaptability and transformability in social–ecological systems. Ecology and society, 9(2).
Wang, N. (1999). Rethinking authenticity in tourism experience. *Annals of tourism research*, *26*(2), 349–370.
WHF (2020). *Bridge to crafts career program*. Retrieved from: https://www.wmf.org/project/bridge-crafts-careers-program

3 Authentication and the authenticating process

This chapter offers an analytical deconstruction of authenticity and the authentication process and identifies its various delineations. It argues that the 'how', in respect to the authentication of authenticity, is also important.

The Ten Commandments letter to the Italian Government, on behalf of the residents and heritage institutions of Venice, said: "We thought we'd take advantage of this last chance to see Venice when it is only for us, alone. It's like having the museum to ourselves. We don't want to go back to that. I want my city to be a real city. Airbnb is like our Covid, it's like a plague, and it turned us into a ghost town" (Horowitz 2020). How authenticity will be authenticated to restore Venice heritage, while accommodating preferred tourist market segments in the post-pandemic era, remains to be seen. In the context of Venice and other places, facing similar dilemmas around the world, several questions arise: Who will be the role players in the authentication process and will this authentication withstand the test of time while securing commercial interests? These questions prompt retrospection of authentication processes pursued to date to authenticate or deauthenticate heritage and for what purpose? This chapter offers a dialectical view on these topics.

Scholars acknowledge that authenticity of heritage resources is a contested and negotiated phenomenon. It can be attached to an aura imbued to the physical attributes of a site or an object (Rickly-Boyd 2012). Existentialist authenticity can be experienced by engaging with that aura. The authenticity discourse and the volatility of its different notions can be best comprehended through the lens of authentication. Insights into the authentication process can unfold how a particular type of authenticity is bestowed on an object or experience. Undeniably this process "involves different actors that (re)construct authenticity for different purposes in terms of prevailing political, economic or cultural conditions" (Su, Song, & Sigley 2019, p. 4). Rather than centering the discussion on what is 'authentic' or 'inauthentic', focus needs to move to the identification of key role players who hold the power to define or shape authenticity (Chhabra, Lee, Zhao, & Scott 2013; Cohen & Cohen 2012; Su et al. 2019; Xie 2011). Several dynamics command the manner in which experts are legitimized and afforded the power to confirm or shape their preferred version of authenticity. In his reflection on authentication of 'ruralness', Frisvolli writes that "the contestation, negotiation and consumption of space involve the interaction of ideas, locality and human practice; Authenticity's social sides, such as power and the moral organization, are often blinded to decide what to display and what to offer for consumption to tourists" (2003, p. 273).

As the authentication perspective continues to ignite interest, critical insights are needed on how heritage institutions and mediating agents authenticate heritage (Chhabra et al. 2013; Wall & Xie 2005; Xie 2011). In this chapter, I offer an analytical deconstruction of authenticity and the authentication process and identify its various delineations. The 'how', in respect to the authentication of authenticity, is also important. This chapter is structured around three areas:

- Deconstruction of authenticity
- Delineation of authentication
- Authentication and its social process

Deconstruction of authenticity

Unarguably, three broadbands delineate the notion of authenticity: objectivist, constructivist, and existentialist. Conceptual bridges, such as theoplacity and a negotiated stance between objectivist and constructivist ideologies, have prompted numerous deliberations in documented literature. Most recent literature unpacks the tension between authenticity and inauthenticity of one's true self, on moral and hypocritical grounds, which also calls for reconciliation between the two. Previous chapter has shared reflexive insights on the pluralist aspects of authenticity. More recent literature has drawn attention to the social-constructivist scrutiny of the authenticity phenomenon and stimulated the much needed attention on the process through which authenticity is authenticated (Chhabra et al. 2013; Cohen & Cohen 2012; Su et al. 2019). This perspective builds on the notion that authenticity in tourism is something that is negotiated; it involves institutionalization and interplay of power and involvement of a multitude of actors, actions/practices, knowledge, and traditions (Frisvolli 2012, p. 274). Cohen and Cohen pose a question in their study: "is there a difference between processes through which objective, as against existential, authenticity is established? Who has the power to endow tourist attractions with authenticity?" (2012, p. 1296). To date, limited studies have deeply and conceptually endeavored to fathom how authenticity is authenticated or conferred on cultural and heritage objects and experiences (Chhabra 2012; Cohen & Cohen 2012; Frisvolli 2013; Su et al. 2019; Xie 2011)). Cohen and Cohen (2012, p. 1296) define authentication as "a process through which something – a role, product, site, object or event – is confirmed as 'original,' 'genuine,' 'real' or 'trustworthy.'"

Delineation of authentication

Three key dimensions of authentication can be noted in documented literature: hot, cool, and mutual. Cohen and Cohen (2012) offer insights on hot and cool authentication. The authors state that the purpose of cool authentication is to freeze and fossilize heritage. This stance is especially evident in the manner objects and artifacts are museumified in museums under the umbrella of cool authentication. Hot authentication, on the other hand, seeks to inspire based on vitality and buoyancy. The authors explain that these concepts offer insights on how objective and existentialist versions of authenticity are legitimized. Cohen and Cohen write that these two types of authentication have different dynamics and they occasionally intersect, coinfluence each other, and sometimes they are positioned on two opposite ends of a continuum. The legitimization of objective and existentialist authenticities can be explained by examining the authentication process. Cohen and Cohen write that "authentication endows an object, site or event with authenticity; it thus involves performativity …, which has been recently introduced into the discourse of authenticity …. However, as we illustrate below, 'performativity' becomes a 'speech act' in the context of cool authentication, whereas in 'hot' authentication it becomes as implied in the use of the term … a constitutive performative process" (2012, p. 1298).

The 'hot' authentication process will continue to evolve "as performative norms continue to be continually enacted to retain their power" (Edensor 2001, p. 62). It is postulated that "the two modes of authentication in some instances mix without tension and can be seen as co-constitutive of the authenticity of a site. An interesting manifestation of such coalescence is the inclination of visitors to 'hotly' authenticate some aspects of otherwise 'coolly' authenticated attractions, such as monuments or museum objects" (Cohen & Cohen 2012, p. 1304). A notable example of cool authentication is offered by Noy (2009) in his study of the Ammunition Hill Museum in Jerusalem. Noy writes that "the 'cool' authentication of the museum is provided by the authorities who choose to locate it on the actual site of the battle for the hill and display in its original documents and other artifacts. But that authentication is supplemented by the 'hot' authentication of the site by way of visitors' emotion-loaded inscriptions in the visitors book" (Noy 2009, p. 121).

In tourist settings, it is highly likely that boundaries between staged representations and the audience become blurred as local communities accept/reiterate the cool authentication process or question the manner in which objects, locations, and occasions are legitimized (Edensor 2001). Tourist attractions are often contested spaces where a single culture can be simultaneously contested, negotiated, and/or consumed (Lacy & Douglass 2002). On politics of authentication, Cohen and Cohen write that:

> it can be argued that it is not a zero-some game but is controversy and disputable. It is subject to 'conflict of interest' and hence it can be

> politically implicated. As pointed out by Bruner, the basic question is not of whether the site is authentic or not but rather it was authenticated by whose authority. The question here is of power. The authors are of the view that power plays a different role in both types of authentication. Politics of authentication are dynamic and differ across the two modes. Who has the power? In the case of cool authentication, the power is vested in experts and custodians and in the case of hot authentication, whereas power in 'hot' authentication … becomes a constitutive performative process (2012, p. 1298).

Similar perspective is shared by Su et al. (2019) in their narrative on reconstructing heritage of Shaolin Temple to increase heritage tourism in China where they discuss the issues associated with the legitimization of heritage and remodeling of the temple landscape. The authors outline the manner in which the Western authorized heritage discourse (AHD) of authentication differs from that of the Chinese AHD. As a case in point, they draw attention to the reconstruction of tangible heritage versus restoration of built heritage as preferred by the Western AHD. It is pointed out that the recent Chinese scholars dismiss the reconstruction initiatives of ancient buildings. From Su et al.: "Liang's restoration principle has no doubt legitimized small-scale reconstruction through restoring partly ruined buildings to their 'original-state' in a certain dynastic style. In this regard, authorized restoration or reconstruction guided by heritage experts is regarded as the legitimate authentication of built heritage. At the same time, if the reconstruction is large-scale or total, in particular if it is for touristic or commercial use, it would be condemned as artificial or fake, and simply seen as a historical theme-park" (2019, p. 4). Table 3.1 frames Su et al.'s explanation of how the temple is authenticated.

It is interesting to note that a decade or so ago, the process of reconstruction and renovation of the temple was considered less authentic because when it was largely reliant on oral narratives/memories of old people (Su et al. 2019). It is deemed to be more authentic now that copied images of authentic martial arts, from Japan, are being used to authenticate expressions and displays (Su et al. 2019). However, this is the management standpoint. The monks and the local residents differ from this perspective and also from each other with regard to perceived authenticity of Shaolin cultural heritage. According to the monks, Chan Buddhism made the Shaolin heritage and culture authentic. They feel that its spirit is still intact and the physical form of the temple is not that important. While the locals also show complete disregard for the tangible authenticity of the buildings, their perception of authenticity appears to be different from the monks. They claim to be the true inheritors of Shaolin martial arts (from the Maoist and early Dengist times) and argue that the current martial art performances, authenticated based on evidence from Japan, are orchestrated in an inauthentic manner. Also, Su et al. report that the local residents express displeasure

Table 3.1 External and internal authentication of the Shaolin Temple

The project-compelled departure of most commercial businesses (of 1980s and 1990s) to make the Songshan Shaolin Scenic Area (SSSA) environment more 'authentic'. The rapid tourism development of the 1980s and 1990s had resulted in an unregulated, one kilometer, commercial street stretching from the scenic zone gate to the actual gate of the Shaolin Temple. The street became crowded with 690 retailers, several private martial arts schools, and numerous enterprises such as tourism businesses and government agencies.
The tourism businesses listened to the views of the local government, the Shaolin monks, and the local residents albeit they became notorious for compromising the 'authenticity' of the Temple. Businesses involving the local residents particularly received negative coverage and were considered 'disharmonious' to the temple setting. These businesses and local residents were given orders to relocate. However, the tourism businesses, which involved the Shaolin monks and the government departments, became more acceptable after they were modified to adapt to the authenticity criteria of the World Heritage Sites.
It was in 2004 that the internal-temple authentication project was launched and it embraced the processes of restoration and reconstruction to authenticate the buildings inside the Shaolin Temple. First, some buildings were restored so that they could look more 'authentic'; second, some disharmonious buildings, which were not synchronized with the temple, were demolished to create a more 'authentic' environment; and third, Qing-dynastic style was used in the reconstruction and repair of some buildings.
Nevertheless, the religious life of the monks was taken into consideration during the authentication of the internal-temple. Some new buildings were constructed to adhere to the contemporary requirements of the religious life of the monks. Some areas were made inaccessible to protect the privacy of the monks. For instance, the dorms and meditation spaces were built in a low-key manner so that the monks could carry on with their religious life, away from the gazing and intrusion of the tourists. As another example, the reconstructed 'Ordination Platform' was designed to meet the religious needs and tourists were not allowed to enter. It is reported by the construction bureau that Renminbi (RMB) 50 million, toward the authentication project, was contributed by the monks.

Source: Su et al. (2019).

because they were compelled to relocate, although the government offered them substantial financial incentives in the form of housing at subsidized rates. They are emotionally attached to the land near the Shaolin Temple and claim that the space belongs to their ancestors. In examining these dissonant and contesting views, it is shared that:

> the external temple authentication reveals the hegemonic influence of the AHD in that the local residents were forced to relocate outside to 'purify' the integral environment of Shaolin heritage. Just as objectivist authenticity focuses on the legitimization and heritage experts, the external-temple authentication demonstrates the determinate role of experts in removing the disharmonious scenes to make a 'clean/tidy' environment more befitting a contemporary world-class tourist

> attraction. The Western AHD focuses on the materialistic heritage with disregard to local voices …., while the SSSA case shows that the Shaolin monks take the semi-role of experts and local users of the built heritages. Clearly, both Shaolin monks and local residents try to claim their legitimacy to 'own' authentic Shaolin heritage (Su et al. 2019, pp. 15–16).

In a nutshell, Su et al. illustrate how authentication can become a politically contested process if important stakeholders hold opposing views. This is an example of dissonant heritage. Clearly some voices are being disregarded by the local government and a preferred version is legitimized to authenticate objective authenticity of the Shaolin Temple. It is claimed that the authentication process is scaled to align with the Western AHD evidenced from focus on a clean environment, displacement of local tourism businesses to pave the way for large-scale tourism businesses, and use of selective historical records to guide the restoration and reconstruction of buildings. The Western-centered concepts are condemned for 'petrifying, museumifying, and decontextualizing' heritage as static and isolated objects, because they promote segregation of heritage from daily life. Obviously, this is a case of conflict between the local and universal as the government is more inclined to showcase the attraction for international audience (DeSoucey, Elliott, & Schmutz 2019).

Another case in point is presented by Zhu (2015). The author reports that the Chinese government legitimizes, in a selective manner, the notion of authenticity for local heritage practices such as prioritizing heritage, conservation, and management. Authentication is a social process that shapes local customs. He presents several influences of authentication on local cultural practices that are shaped by a combination of local (by the Chinese government) and global values (inscribed by the Venice Charter). Since authenticity has several connotations, the best approach is to tie the notion of authentication with different socio-constructivist processes through which authenticity is recognized and defined. Lacy and Douglas argue that tourist sites are spaces "within which multiple interpretations of a single ostensible culture can be negotiated, contested and consumed" (2002, p. 7).

Debunking the hot and cool notions of authentication, Frisvolli argues that they fall short of actual application of the authenticity concept and are limited to authoritarian endorsement or subjective assessment guided by performative practices that are based on perceptions and beliefs. Authenticating rurality has a conceptual reference point as 'rural space is socially produced' (2012, p. 275). And, according to the author, Cohen and Cohen's work lacks an analytical approach. Of more concern is the need for addressing issues such as determining the foundations/essence of ruralness in a tourism product and identifying factors that are likely to influence, bestow continuity, and help with reconstruction and reproduction.

To fill this gap, Frisvolli suggests a threefold approach, a conceptually guided analytical deconstruction of authentication with three overarching domains: rural showcasing (representations), rural neighborhoods/locales, and their attributes and rural lifetimes/existences. Frisvolli argues that these three elements are interdependent on each other and are crucial in understanding the process of rural authentication, especially in the context of tourism:

> First two are the material dimensions of rural space which can be translated in touristic terms to the 'toured objects' (or activities) and their material context (e.g. elements in the surrounding landscape or present at the tourist site). These dimensions refer to elements introduced deliberately to the tourism product and to elements beyond the control of the tourist entrepreneur/host. The next dimension on rural existences refers to people's subjective reproduction of the rural through everyday life practices which can evolve and be varied. In terms of tourism, this also relates to tourists' and tourist hosts' performances and their physical connection with the tourism resource. This dimension represents the potential to include tourists' and tourist hosts'/entrepreneurs' feelings and assessments of a tourism product/experience. I argue that such an approach to authentication has the potential to analytically recognize how such sentiments feed into the tourism product and the rurality they are perceived to represent (2013, p. 277).

In the aforementioned, two elements of relevance are stressed: rural space's three dimensions and the structural coherence of rural space. Structural coherence refers to power and ability of a tourism product to showcase itself as authentic; it is often a subjective value bestowed on rurality or a tourism product conferred by the tourist, host, or the institution where the product is housed. Frisvolli argues that it is important to understand which view of authenticity (in the context of his focus-authentic ruralness) is correct, how and why. This implies that the authentication process of a space or a product spans different steps such as staging, designing, and grooming. He suggests a Halfacreean perspective to explore this process of authentication. His proposed framework is focused on deconstructing authentication and includes six core intertwined elements: production, hardware, intra-coherency, consumption, software, and inter-coherency (Table 3.2). Each component represents an investigative feature and each dimension helps to dig deep into the social processes that shape authentication.

Although Frisvolli claims that Cohen and Cohen's hot and cool authentication are overtly simplistic, few similarities can be observed between the two studies. While Frisvolli's proposed framework (Table 3.2) captures the dynamic process of authentication and the settings within which different notions prevail and why they do so, it fails to highlight the role of different

Table 3.2 Deconstructing authentication of authentic ruralness

Dimensions
Production
Production process of the tourism product needs to be investigated. Thus, it urges analytical questions such as: What is the tourism product? To what extant do the hosts believe that their rural tourism product embodies a real ruralness? To what extent do the producers see authenticity as a product? And where in the product is its authenticity thought to reside (i.e. artefacts, practice, traditions, etc.)?
Consumption
This box refers to the touristic moment of experiencing the product and what that moment reveals: what is really consumed? Is it a tourism product's higher order function (i.e. a meal as a means to still ones hunger)? To analytically cover this, the second box comments that the analysis asks: how central to the purpose of consuming a particular tourism product is its ruralness and how central is consumption of the 'authentic (i.e. the real)'?
Hardware
This box refers to the elements utilized to convey ruralness in the tourism product. These may be material elements such as farm buildings, animals, and landscapes, but also nonmaterial elements such as rural practices. 'Hardware' refers to the visual, to the 'hands on' artefacts, and actors constituting the tourism product and its surrounding countryside. This box urges the exploration of 'what is mobilised and employed by tourist hosts in producing the offered tourism product'? And 'what is included in the consumption by the consuming tourists'?
Software
This box refers to the cultural lens (i.e. notions of rural representations) and personal input (i.e. previous experiences, assessments, beliefs, etc.) layered onto the hardware in acts of production and consumption by tourist hosts and tourists alike. In other words, software is all those things involved in moving beyond the physical character of the 'hardware'. This box aims to capture the yardstick by which tourist products are assessed. It is important to note that 'software' not only applies to the tourists consuming a rural tourism product, but also to hosts that are constructing/grooming ruralness into a tourism product, as it is conceptually referring to the 'template' by means of which the product is produced/groomed. This conceptual box provides impetus to explore the popular myths of the rural (i.e. rural representations) and their integration with the production and consumption processes of a tourism product.. The last two boxes reflect on the political side of authentication.
Intra-coherency
Warrants an investigation of the relationship between the rural tourism product and its rural space, which could be quite simply defined as what is seen, heard, and otherwise sensed (i.e. experienced) while consuming a rural tourism enterprise's product and the significance attributed to these aspects (i.e. their meaning). This dimension parallels Halfacree's (2007) notion of structural coherence, and the key question to address here is to what extent do the different elements of a given rural product and its surrounding space tell the same story: Is the rural narrative of the tourism product internalized in the elements present, or is it undermined? An important aspect of rural as socially produced is the idea that representations are formed, sustained and changed in a discourse with other representational meanings (Frisvolli 2012; Halfacree 2007).
Inter-coherency
This last box encourages an inquiry into this, urging that the linkages of rurality are explored. This conceptual box sets out to ask: how does the tourism product investigated relate to the larger discourses involving countryside and regional issues and to what extent is this reflected in the tourist's consumption of the products?

Source: Frisvolli (2013, pp. 279–280).

players/stakeholders. Nevertheless, the author furthers the discourse and stresses on the need for a mutual form of authentication which is, in fact, a:

> multifaceted mesh of materiality, social representations, political discourses, practices, and performativity. A common point of reference is noted between visitors and the hosts. A set of representations is used and consumption is compared to the selected elements. These elements differ across stakeholders and hold a potential of generating a disharmonious authenticity especially when they get entwined with the political process of authentication. Heritage tourism can play a mentoring role in conferring authentication to heritage spaces and confirm selected elements to be true representations of 'heritageness.' Comprehensive data is needed, such as information to be gathered on all factors (with special stress on representations, lifestyles and local settings) of heritage in addition to the challenges posed by the six identified domains, if the authentication process has to be examined in a complete manner (Frisvolli 2012, p. 294).

Moving forward, a recent study offers a reciprocal standpoint. Mutual authentication by guests and hosts, in the context of intangible cultural heritage (ICH), is examined by Khanom, Moyle, Scott, and Kennelly. Host perspectives of authentication cannot be overlooked because unquestionably, the host community has constructed, maintained, and transmitted ICH across many generations (Zhu et al. 2015). While the role of host communities in the authentication process is crucial (Khanom, Moyle, Scott, & Kennelly 2019; UNESCO 2011), demand and institutional perspectives are also important to facilitate sustainable development of tourism and safeguard the ICH (Chhabra et al. 2013; Cohen & Cohen 2012) from exploitation. Empowerment in the selection of authenticating markers, for showcasing of heritage, can contribute to the conservation of ICH (Alexander 2009; Khanom et al. 2019) and authenticate the path for sustainable heritage tourism.

Authentication is also considered from the standpoint of integrity. Zhao, Campisi, and Kundur write that, "authentication is the service of ensuring whether a given block of data has integrity (i.e. the associated content has not been modified, and is from the legitimate sender" (2004, p. 430). The authors share rhetorically that:

> authentication is traditionally confirmed through mechanisms that involve message authentication codes (MAC) and digital signatures known as hard authenticators. In hard authentication, a MAC (also known as a message digest) or digital signature of the data to protect, called an authenticator, is created at the source and transmitted with the data. At the receiver, the authenticator is verified using the received data to deduce if the received information is in fact unmodified and from the alleged sender (2004, p. 430).

From the foregoing, authentication has been examined in a variety of contexts such as ruralness, ICH, and world heritage site (WHS). All recognize the "need to gain an insight into the overall authentication process and to understand the role of different authenticating agents and markers in defining/endorsing authenticity of a product, site or a destination" (Chhabra et al. 2013, p. 146). Ethnic cuisines also offer a notable context because "ethnicity is conveyed in a purpose-driven staged manner in spaces beyond its place of origin, for instance, much ethnicity, in a US setting, is made real through transactions between ethnic entrepreneurs and the public via mercantile strategies" (Lu & Fine 1995, p. 535). Sims (2009) writes that engaging experiences, employing cultural traditions in a meaningful manner, can facilitate authentic connections between the original locations and tourism products. In the last decade, numerous studies have examined authentication of heritage experience associated with ethnic cuisines. For example, in the case of ethnic cuisines abroad, the authentication process can refer to the manner by which restaurant managers endorse authenticity in ethnic or traditional food items. It uses predetermined authorized criteria and authenticating markers to localize a product in a distinct manner. In the context of ethnic cuisines, "degrees of portrayed 'Indianness,' 'Thainess,' and 'Hawaiianess' illustrate efforts of ethnic restaurants and other cultural institutions to carry forward traditional elements of culture and provide a sense of 'othered' continuity to contemporary audience" (Chhabra et al. 2013, p. 155). This 'othering' illustrates the manner in which the other culture is authenticated so that it can be celebrated and admired for its distinction.

Authentication can be referred to as a process that facilitates foodservice experiences (Mkono 2012). According to Lu and Fine (1995), it refers to a point of negotiation that embraces both what is constant and what is fluid. The key purpose of adaptation, in the context of ethnic food, is to make it palatable for economic numerations. Along similar lines, Edensor describes it as a process where multiple meanings are conferred by various agents to redefine and connect to the needs of different markets. Chhabra et al. (2013) examine the authentication of Indian cuisine abroad using a set of authenticating markers. The authors report poor efforts on the part of the restaurants to communicate ethnic 'Indianness' by using authenticating markers such as tangible décor etc. It is suggested that a broad range of authenticating agents need to be identified and used to convey 'Indianness' and add to the ethnic appeal.

Authentication as a social process

Lamont (2014) makes an effort to examine social processes that ascribe authenticity to tourism offerings in the context of sports tourism. He examines social practices of sports tourists in the context of a commercial tour package that confirm and exaggerate the French Alps as an 'authentic' tourist space. Social processes through which authenticity is authenticated continue to miss scholarly attention. Lamont's paper explores dialectics of entrepreneurship and cultural consumption in the tourist enclave of 'Foreigner's Lane' in Dali, Yunnan Province, PR China. He focuses on ethnic identities and their orchestration to attract tourists. Lamont uses the performance metaphor to conceptualize tourism as a carefully staged act and examines it using the authentification lens. Local entrepreneurs have constructed 'exotic otherness' to align with the preconceptions and demands of backpacker travelers.

Gao and Bischoping (2019) examine cultural politics of authentication and fakery in the manner the China's Lei Feng legend is curated by the State and the Western media. Mixed response of different audiences evokes deliberations on the power interplay for legitimizing authenticity. The authors offer insights on how narrative of a national figure is authenticated by China and by another (western) nation:

> By highlighting the sharp contrast between the views of the Chinese advocates of Lei Feng and of the western sceptics, we have demonstrated that besides asking who authenticates and how, we must scrutinize the epistemic condition that legitimizes one method of authentication while dismissing another. We have first shown that Chinese research participants view what sceptics would call 'adulterated' photographs as simultaneously a form of evidence and a pedagogical tool that the state can justifiably curate for the social good. Meanwhile sceptics selectively focus on evidential flaws in the Chinese Communist narratives while leaving western democracies' narrative inconstancies unquestioned. Their discourse about authenticity rests on the stereotypical association of propaganda with Communism. We have shown how this story has been differentially decoded according to frames predetermined by their political implications (Gao & Bischoping 2019, p. 12).

Different scholars have deliberated on the shift of authenticity to authentication for the purpose of gathering insights into the social/political processes through which exclusiveness and genuineness is confirmed or legitimized. To a great extent, the process through which authenticity is endorsed, rationalized, or negotiated is shaped by the privileged position of the key stakeholder/s whether it is the government, a tourism business, tourist, or the ethnic community (Xie 2011). Ateljevic and Doorne (2005) use a dialogical perspective to unpack the process through which authentication is mediated between suppliers and consumers. Both studies take a monolithic view of the social process, instead of delineating them into pluralistic frameworks, to acknowledge the role of important players and identify the unique/harmonious aspects of heritage settings. Conflicting claims by key stakeholders are commonplace and power dynamics often shape the legitimization of claims.

Closing comments

Authentication still remains a fledgling area of research and an unchartered territory. Different markers and perceptions of plausible reality exist. This chapter offers several scenarios to illustrate that authentication is a multifaceted social process shaped by power relationships. Different types of authentication processes are noted, shaped by location specific socio-constructivist environments: hot, cool, and mutual (interconnected between hosts and guests). These are supported by tangible and intangible markers and occurrences to legitimize and communicate objective versions of authenticity. Undeniably, objective authenticity remains the enshrined reference point for heritage tourism institutions that are keen to use heritage as a sustainable commodity. Mutual authentication of negotiated authenticity, particularly, has perpetuated the attention of both, the academicians and the practitioners. Khanom et al. (2019), while enshrining ICH, support the process of mutual authentication between the guests and the host communities. The authors argue that "mutual authentication of ICH and community empowerment could help to prevent the commodification of ICH as well as strengthen the community's role in ICH based tourism" (2019, p. 5).

All scenarios presented in this chapter unarguably revolve around negotiated authenticity, holding objective authenticity as its mirror, in an effort to stay close to the core essence of the heritage resource. Authentication effort to "authenticate an object is an achievement that requires persistent maintenance; this implies that the suspicion of fakery will perpetually continue to hover. Social actors' choice can hinge on the framing's political potentiality. Therefore, if social debates persevere in referring simplistically to authenticity as fakeness without recognizing the complexities of their making and decoding, then debate will do more to entrench ideological divides than to lead toward a truth-based consensus" (Gao & Bischoping 2019, p. 13). In other words, a consensus or genuinely negotiated stance requires an in-depth probe into the authentication process. Both, political and socio-constructivist perspectives, are crucial in this regard. Undeniably, an authentic experience predicates on harmonious trade-offs between tourism entrepreneurs, government, and other key stakeholders who act as mediators between hosts and guests (Chhabra 2012). A 'double-bind' authentication movement between visitors and institutions, aiming to retain objectively authentic elements, commands attention in future studies. Conceptual content needs to focus on the operationalization of the process, by identifying and examining relevant human agencies that have the authority to authenticate, and deconstruct the power dynamics behind this authority.

Different alienated forms of authentication (such as hot, cool, mutual, and double-bind) lay bare some harsh truths associated with the manner these processes are developed and legitimized. Scrutiny unpacks slanted and dissonant authentications by a handful of mediating agencies and absence of local voices in commodified heritage expressions in several heritage settings.

According to Anderson and Knee, the COVID-19 crisis calls for the need for "shifting priorities toward collective social responsibilities, mutual interdependencies, shared resources, and collaborative approaches" (2020, p. 3). However, collective social processes run the risk of compromising the uniqueness of heritage. Therefore, shifts toward a collective identity run the risk of melting all heritages to develop a shared heritage. On a positive note:

Unarguably, today the technology has proved to be an asset for transmitting and sharing past heritages and their authenticities at homes. There are plentiful ways to showcase them in the form of digital narratives and visuals in an innovative and immersive manner. In the pre-pandemic era, negotiated experiences rarely happened in isolation from the physical spaces. It has brought communities together in digital spaces although in the long run, it will not be able to authenticate the experience of physical spaces. Nevertheless, the pandemic has opened opportunities for retrospection and reformed authentication processes that can focus attention on unexplored prospects to discover/protect ignored or forgotten pieces of past. As a case in point, in the dream of a new Venice by local residents calls for realistically negotiated scenarios. Enshrining Venice and keeping mass and other serious tourists with lower spending propensity, away will call for a reformation of public policies and adequate fiscal support from the government.

References

Alexander, N. (2009). Brand authentication: Creating and maintaining brand auras. *European Journal of Marketing*, *43*(3/4), 551–562.

Ateljevic, I., & Doorne, S. (2005). Dialectics of authentication: Performing 'exotic otherness' in a backpacker enclave of Dali, China. *Journal of Tourism and Cultural Change*, *3*(1), 1–17.

Chhabra, D. (2012). Authenticity of the objectively authentic. *Annals of Tourism Research*, *39*(1), 499–502.

Chhabra, D., Lee, W., Zhao, S., & Scott, K. (2013). Marketing of ethnic food experiences: Authentication of Indian cuisine abroad. *Journal of Heritage Tourism*, *8*(2–3), 145–157.

Cohen, E., & Cohen, S. (2012). Authentication: Hot and cool. *Annals of Tourism Research*, *39*(3), 1295–1314.

DeSoucey, M., Elliott, M. A., & Schmutz, V. (2019). Rationalized authenticity and the transnational spread of intangible cultural heritage. *Poetics*, *75*, 101332.

Edensor, T. (2001). Performing tourism, staging tourism: (Re) producing tourist space and practice. *Tourist Studies*, *1*(1), 59–81.

Frisvolli, S. (2012). Power in the production of spaces transformed by rural tourism. *Journal of Rural Studies*, *28*(4), 447–457.

Frisvolli, S. (2013). Conceptualising authentication of ruralness. *Annals of Tourism Research*, *43*, 272–296.

Gao, Z., & Bischoping, K. (2019). The communist hero and the April Fool's joke: The cultural politics of authentication and fakery. *Social Anthropology*, *0*(0), 1–17. doi: 10.1111/1469.8676.12593.

Halfacree, K. (2007). Trial by space for a 'radical rural': Introducing alternative localities, representations and lives. *Journal of rural studies*, *23*(2), 125–141.

Horowitz, J. (2020). Venice Glimpses a Future With Fewer Tourists, and Likes What It Sees. Retrieved on May 25, 2020 from: https://www.nytimes.com/2020/06/03/world/europe/coronavirus-venice-tourists.html

Khanom, S., Moyle, B., Scott, N., & Kennelly, M. (2019). Host–guest authentication of intangible cultural heritage: A literature review and conceptual model. *Journal of Heritage Tourism*, 1–13.

Lacy, J. A., & Douglas, W. A. (2002). Beyond authenticity. The meanings and uses of cultural tourism. *Tourist Studies*, *2*(1), 5–21.

Lamont, M. (2014). Authentication in sports tourism. *Annals of Tourism Research*, *45*, 1–17.

Lu, S., & Fine, G. A. (1995). The presentation of ethnic authenticity: Chinese food as a social accomplishment. *The Sociological Quarterly*, *36*(3), 535–553.

Mkono, M. (2012). A netnographic examination of constructive authenticity in Victoria Falls tourist (restaurant) experiences. *International Journal of Hospitality Management*, *31*(2), 387–394.

Noy, C. (2009). The politics of authenticity in a national heritage site in Israel. *Qualitative Sociology Review*, *5*(1), 112–129.

Rickly-Boyd, J. M. (2012). Authenticity & aura: A Benjaminian approach to tourism. *Annals of Tourism Research*, *39*(1), 269–289.

Sims, R. (2009). Food, place and authenticity: Local food and the sustainable tourism experience. *Journal of Sustainable Tourism*, *17*(3), 321–336.

Su, X., Song, C., & Sigley, G. (2019). The uses of reconstructing heritage in China: Tourism, heritage authorization, and spatial transformation of the Shaolin temple. *Sustainability*, *11*, 4111 doi: 10.3390/su11020411.

UNESCO. (2011). What is intangible cultural heritage? *UNESCO*. Retrieved 2 November from https://ich.unesco.org/doc/src/01851-EN.pdf

Wall, G., & Xie, P. (2005). Authenticating ethnic tourism: Li Dancers' perspectives. *Asia Pacific Journal of Tourism Research*, *10*(1), 1–21.

Xie, P. (2011). *Authentication of ethnic tourism*. Bristol: Channel View Publications.

Zhao, Y., Campisi, P., & Kundur, D. (2004). Dual domain watermarking for authentication and compression of cultural heritage images. *IEEE Transactions of Image Processing*, *13*(3), 430–448.

Zhu, Y. (2015). Cultural effects of authenticity: Contested heritage practices in China. *International Journal of Heritage Studies*, *21*(6), 594–608.

4 Information communication technology and digitalization of heritage tourism

This chapter offers a discourse on how authenticity is mediated and authenticated in the digital realm. First, the benefits of Information and Communication Technology (ICT) are celebrated and then its drawbacks and challenges are pinpointed. The chapter closes with deliberations, on inspiring and sustainably innovative examples and possibilities, for cultural and heritage institutions to digitally engage their audience in their homes.

Digitalization of heritage can offer numerous benefits if the process remains true to its 'to conserve heritage' vision, plans, and manages authenticity in a harmonious, equitable, and sustainable manner. Undeniably, the COVID-19 has paused the heritage tourism industry. But, by the same token, it has facilitated retrospection of the manner cultural/heritage is showcased for gazing and sale. The pandemic has impacted heritage sites across the world, but the magnitude of impact varies. As evidenced in a behind-the-stage narrative from the World Monuments Fund (WMF):

> Every day at World Monuments Fund, we think about the significance of our heritage in contemporary life, the importance of cultural exchange, the enrichment that travel affords, and the impact it all has on local communities around the world. The current situation has turned much of this on its head. Now more than ever, we are thinking about what our work means to the people with whom we partner globally. At WMF, we are closely monitoring the situation in each of the 29 countries where we operate and basing decisions on the advice of local authorities as well as the World Health Organization (WMF 2020).

A radical shift is evident in the manner the heritage tourism industry is showcasing and marketing its tangible and intangible heritage to heritage audiences. Information and Communication Technology (ICT) has transformed the way services are delivered/obtained and expended (Chhabra 2015; Ukpabi & Karjaluoto 2017) in the heritage tourism environment. According to Arsul, Lumenta, and Sugiarso (2015), e-Tourism relies on ICT to deliver a variety of services to consumers and holds tremendous potential to extend outreach simultaneously to a vast audience. Due to ease of access, reduced costs, globally disseminated opportunities, and customized products/services more and more travelers are looking for unique digital experiences. Boo writes that: Most ancient artifacts and sites are vulnerable to corrosion because of age. Furthermore, unpredictable natural and manmade catastrophes can damage them. In the face of such disasters, there has been a growing voice and movement to digitize cultural heritage to preserve them in the face of potential mishaps such as climate change, natural calamity, weak policy or deficient infrastructure. Another function of a more digital landscape can be to ensure the accessibility and usability of the digitalized data. Institutions such as the Smithsonian – have opened their collections to the public for viewing, albeit many of them continue to struggle with visibility and broad access issues (Boo 2020).

The significance of ICT has become more pronounced today, in the pandemic times, when physical travel is paused or restricted. Most heritage institutions are offering virtual tours and experiences, as they wait for the corona virus threat to subside. This chapter offers a discourse on how authenticity is mediated and authenticated in the digital realm. First, the benefits of ICT are celebrated and then its drawbacks and challenges are pinpointed. The chapter closes with deliberations, on inspiring and sustainably innovative examples and possibilities, for cultural and heritage institutions to digitally engage their audience from their homes and backyards. The key highlights of this chapter are:

- Meaning and benefits of digitalization
- Digitalization of cultural and heritage resources
- Digital authentication of negotiated authenticity
- Sustainable digitalization of heritage tourism

Meaning and benefits of digitalization

Undoubtedly, digital innovation has become crucial to the success and competitiveness of all heritage institutions and their stakeholders (Konstantinou 2016; Ukpabi & Karjaluoto 2017). Top activities searched, on the world wide web, by consumers are associated with sightseeing and culture and heritage (Chow & Murphy 2011). Cultural and heritage showcasing, on destination websites, is a marketing tool to attract consumer attention. Web 2.0 and web 3.0 are the contours of e-tourism. Web 2.0 brought with it a wide assortment of electronic applications (such as social media, review sites, blogs, interactive websites, photo and video sharing platforms) that facilitated interactions between users and heritage agencies. Unprecedented popularity of mobile technology has also extensively fashioned consumer demand and stimulated recognition of heritage tourism products and services.

According to Popescu, Nicolae, and Pavel (2015), digital applications have helped to expand outreach by offering user-friendly interfaces while at the same time they facilitate stakeholder partnerships with lower investment costs. They have transformed the heritage tourism value chain and augmented strategic relationships between tourism organizations. This has been further boosted by the invention of smartphones or Web. Furthermore, new social media platforms (for instance, Google+ and Instagram) or interactive websites (for instance, TripAdvisor and Yelp) have eased connectivity by offering user friendly podiums for sharing views and experiences (Díaz-Andreu 2017). Websites such as booking.com and skuscanner.com enable a high number of interactions and online transactions (Popescu et al. 2015). In the context of social media, Popescu et al. write that:

> if judiciously managed, social media is a very good tool for awareness and interaction. Having a great return of investment rate, it can generate thousands of shares for a single post. Regarding on-page website management, conversion rate optimization represents a good solution for increasing online sales. The general design and the content provided should always be tested for improvement in order to continuously adapt and optimize the website. In tourism and hospitality industry, user generated social validation has a noticeable impact on consumer behavior. (Popescu et al. 2015, p. 167)

Heritage institutions have often lagged behind in technological initiatives partly because of lack of knowledge/expertise and partially because of lack of resources or budgetary constraints. The manner in which small- and mid-size enterprises (SMEs) and nonprofit organizations embrace technology is impacted by numerous factors such as environmental pressure, budgetary constraints, industry barriers, and training constraints. Furthermore, several game changers are likely to emerge in the post-pandemic future that will unravel new ways of travelling and experiencing heritage; in fact they will shape the entire travel life cycle experience of heritage tourists, starting from decision-making and consumption to post-purchase behavior (Kotler et al. 2017; Ukpabi & Karjaluoto 2017). On a positive note, Natale and Piccininno (2018) point out that cultural institutions are demonstrating more keenness to offer unlimited access to cultural repositories. What is needed is a uniform digital heritage plan that is a win-win game plan for all stakeholders of cultural/heritage tourism. A network infrastructure, with an ability to optimize the use of mobile technology, can facilitate effective use of cultural/heritage assets during and post-pandemic times. In line with this view, Natale and Piccininno (2018) stress on:

- Supporting the digitalization of cultural heritage as a strategic agenda of tourism promotion;
- Encouraging the intelligent use of open-source data to offer innovative tourism services that are based on the cultural heritage of individual institutions and the local area as a whole;
- Facilitating collaboration among all stakeholders in the tourism supply chain to stimulate synergies that are beneficial for all key role players

such as cultural institutions, the private sector, creative industries, and consumers;

- Facilitating tourism initiatives and services that focus on the real needs of users, optimizing the resources offered by the web and digital technologies to enhance the tourist experience before, during, and after the visit.

In a nutshell, digitalization can therefore be described as "a general term for all necessary steps to create and sustain digital resources" (Preuss 2016, p. 8). Digital presentation is the readily available information that employs "search engines or individual websites. The information itself exists in very different formats, such as texts, audio files, images or video files, as well as corresponding metadata in machine-readable exchange formats" (Preuss 2016, p. 2). Digitizing is a process that combines electricity with the hardware and software of a computer. Meaningful digitizing comprises of several elements:

1 Mechanical typing with a keyboard;
2 Creating an electric impulse which is transformed into a combination of 0 and 1;
3 Interpretation into a sign for a letter;
4 Being processed by software into a font and integrated into a file; and
5 Translation into graphical interpretation to put on display (Preuss 2016, p. 8).

Digitalization of culture/heritage

Within the realm of heritage, He, Ma, and Zhang note that recent digital information technologies have offered significant means for "heritage recognition, protection, presentation, and communication" that hold the ability to resolve many contemporary challenges related to the preservation of historic monuments (2017, p. 335). He et al. present a threefold meaning of digital heritage:

1 Unique resources and information with long-term value and significance produced by digital means (i.e. "digital born resources of lasting value");
2 Property rights of resource in the virtual space (i.e. "inherited virtual property"); and
3 Utilization/integration of "cultural heritage" and "digitalization" (i.e. cultural heritage + digitalization) (2018, p. 335).

UNESCO refers to digital heritage as materials, generated by computers, that contain a long-term value and stresses that it should be sustained for future generations. Based on the criteria set by a German national research funding organization, Preuss identifies five different aspects of digital cultural heritage:

1 Content specification
2 Digital cataloging
3 Digital presentation
4 Digital backdrop
5 Digital preservation

Table 4.1 illustrates how digitalization can be interactively employed to communicate/share information related to art. According to Rullani (2010), all heritage experiences are reliant on three criteria: (a) bottom-line production, (b) the digital form, and (c) shared subscription. The first benchmark offers, an open innovative platform for emerging artworks at reasonable costs, for artists as well as consumers. Since consumers are involved from the beginning, they can play an active role in the production, distribution, and promotion of the artwork. The digital criteria enable broader sharing of culturally constructed and art-based resources. The shared contribution benchmark enables viewing and experience (emotionally, cognitively, and

Table 4.1 Bottom-up examples

A website (Produzionadalbasso) was developed in Milan in 2004 by two brothers. It was designed to serve as an independent platform to offer complimentary space to artists to assist with independent productions or coproductions. A 'bottoms-up' approach guided the entire initiative to encourage a traditional (bottoms-up) path for the selection of art ideas and help with market access. Website is open to host any art/heritage project and assists in exploring and lobbying for funding opportunities. Produzionadalbasso does not earn any revenue from successful projects neither does it seek for copyrights on the products hosted and marketed on the website. The system relies on individual participation and offers opportunities to the audience to coproduce products with the artists. It offers a copy of the final product and permits access to previews, forums, and blogs to promote debate and interchange of ideas. Users are given a central role in the production chain and can enjoy a dual status of consumers and producers.
The most captivating experiences of 'production from the bottom' include the *Permanent Scenical Experiences* and those of a group of artists from Sicily (Attraversamente). This group had walked through Sicily over a distance of 700 kilometers and throughout their journey, gathered and recognized/celebrated novel artworks, interviews, and performances.
Outside the country, independent cultural production is presented in two other digital podiums. The first platform was invented in USA and has been employed by more than 100 countries across the globe and its website hosts a broad catalog of project areas such as art, music, gaming, film, writing, technology, photography, invention, food, political, education, community, and performing arts. The second platform specializes in independent film industry and is connected with the International Film Festival at Rotterdam in Amsterdam.

Source: Calcagno and Zabatino (2010)

physically) of an object or event replica, in an immersive manner, thereby facilitating a creative and interactive dialogue between the users and the artists (Calcagno & Zabatino 2010). A bottoms-up example is shared presented in Table 4.1. Natale and Piccininno (2018) suggest that the novel heritage tourism models should be inclusive of:

- Potential tourism offerings such as inclusive packages, which are not easy to cancel or resell and can be complemented by a central online service;
- Mutually beneficial digital products produced as a result of coordination between the private and public heritage sectors;
- New employment opportunities, both within and outside the cultural institutions;
- Impact assessment of online digital services for heritage tourism; and
- Accessible data to enable use/reuse of cultural heritage resources.

Numerous other innovative tools also hold tremendous potential for the heritage tourism industry such as multimodel applications (combing virtual reality [VR] and augmented reality [AR]), gamification, robotics, and artificial intelligence. While numerous examples of multimodal applications and gamification embraced by the cultural heritage institutions exist, robotics and artificial intelligence appear less tapped by the cultural heritage sectors.

Multimodal applications

Next, a generic multimodal environment can facilitate the ability to customize digital heritage interfaces and environments. An important benefit for the users lies in the ability to execute different types of undertakings and the option to switch between numerous reality environments. Classic examples include mixed virtual and AR platforms and immersive systems:

- User Interfaces – these can be non-conventional and can be tailored to a specific setting. Traditional examples include two dimensional (2D) interaction devices such as mouses, keyboards, windows, menus, and icons. Design and interface play a key role to generate interactive and engaging experiences. Simple adaption of interaction styles to three dimensional (3D) will not work as the modification needs to resonate with specific settings/scenarios. Hybrid interfaces have to be designed in an adaptive manner.
- Virtual Environments – virtual museums offer easy access, remote learning, and flexibility. Liarokapis, Petridis, Andrews, and de Freitas (2017) present the example of 'Virtual Ancient Egypt' that offers digital recreation of ancient artefacts to users. A battle in a digital backdrop can be reconstructed to offer the users an opportunity to move around and view from different locations and levels. Another effective

example is the use of a 'personal digital assistant' (PDA) for onsite visitors. This is a better option than the use of audio/visual prerecorded guides. Furthermore, an information retrieval service (with the help of chatbot and speech recognition technologies) can be made easily accessible through spoken language interaction thereby facilitating natural interaction with the user.

- Noteworthy examples of virtual applications in historic settings also include the 3DMURALE project and the Ename project: The 3DMURALE is a 3D multimedia tool that has the ability to record, reconstruct, encode, and visualize archeological ruins such as the ancient city of Sagalassos in Turkey (Cosmas et al. 2001).
- The Ename project uses a nonintrusive interpretation system (Pletinckx, Callebaut, Killebrew, & Silberman 2000) to transform archeological sites into open-air museums: (1) Ename is a picturesque village in the Flemish Ardennes in Belgium, close to the historical city of Oudenaarde. A closer look reveals that it has an archeological park, a church with an intriguing past, a nature reserve, a museum, and a heritage center. Ename has several stories to tell about its past and about what it is today; (2) One story tells the history of the abbey of Ename, founded in 1063 and dismantled in the last years of the eighteenth century. In this blog, we not only tell this story, but we also show how we visualize the rise and fall of this abbey through virtual reconstruction by turning historical texts and archeological research into appealing and state-of-the-art 3D visualizations (Liarokapis et al. 2017, p. 381).

Another famous example is "Walk Through Ancient Olympia" (from the Foundation of the Hellenic World 2019) where an onsite visitor can learn about the ancient games by interacting with non-player characters (NPC) such as athletes in the pentathlon (athletic contest featuring multiple competition events) (Anderson et al. 2001). Many museums use AR applications in both indoor and outdoor settings, where a "mix of real and virtual content is used with devices such as mobile phones or head-mounted displays (HMD), and smart glasses" (Jung & Dieck 2017, p. 111). Furthermore, overlaying digital content over the real setting can offer numerous interactive opportunities. For instance, historical buildings can become alive by reenacting old events and transmitting a story in an engaging manner (Gervautz & Schmalsteig 2012). Liarokapis et al. (2017) share popular examples such as ARCHEOGUIDE (interactive guide to facilitate visual aspects of an archeological site) and LIFEPLUS (offering a high degree of actual interactive immersed experience, which make use of 3D simulations of virtual beings [humans, animals, and plants]) in real settings. The ARCO system constructs tangible platforms that enable visualization of museum artefacts in an interactive table-top AR setting where broken pieces can be assembled and visualized in their original form.

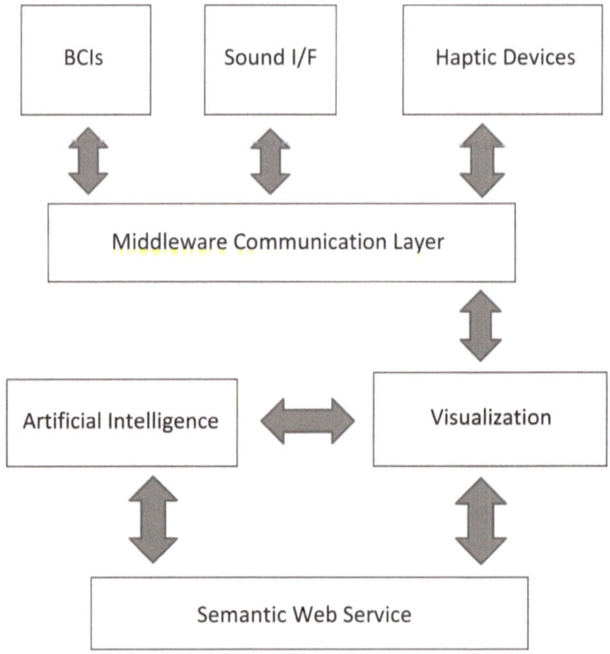

Figure 4.1 Multimodal reality interface.

Source: Liarokapis et al. (2017, p. 385)

Figure 4.1 presents an illustration of how innovative interactive content can be developed by mixing AR, VR, and web imaging. The figure and its description demonstrate that interactive and visual technologies can be used in a multimodal reality system by employing multiple interactive methods such as brain-computer interface interaction; natural speech interaction; and interaction with haptic devices. The haptic interfaces are:

> designed to communicate through the subtle channels of touch-senses that relate to the tactile feature of a virtual object. Additionally, the system supports interaction with several standard input devices and interactive sensors such as the Spacemouse, the Spacepilot, Isense Inertia cube etc. The 3D visualization and haptic interaction can also be apprehended through this architecture on a web browser, virtual gallery environment, person using a physical replica of an artefact or a haptic device to help control and explore the virtual artefact. A Web3D sphere and an AR sphere can be loosely integrated to develop a visual digital heritage system. The Web3D domain can carry out proactive/applied virtual learning through the incorporation of a physical boundary into a virtual environment. (Liarokapis et al. 2017, p. 384)

Haptic devices hold tremendous potential to offer immersive and interactive experiences and their utility in enhancing onsite physical interactions continues to be explored. The ultimate aim of such devices is to optimize visitor experience by stimulating all five senses. Another emerging ICT trend is gamification of heritage. In fact, games have become a powerful medium and hold tremendous potential to motivate a wide range of target markets.

Gamification

Deterding et al. refer to gamification as an "application of game elements and digital game design techniques to non-game problems, such as business and social impact challenges" (quoted in Liarokapis et al. 2017, p. 374). Today, the gaming industry is the most stimulating outcome of technology and gaming is used in websites to enhance consumer encounters, generate loyalty, and brand familiarity. Interactive digital heritage exhibits can be created by combining technologies to provide a multisensory and AR experience. Georgopoulos, Kontogianni, Koutsaftis, and Skamantzari (2017) illustrate how games, especially serious games, can be used by cultural/heritage institutions. Such games facilitate interactive learning experiences by offering virtual tours, encouraging questions about heritage pursuits, items, and shrines. Notable examples include:

- The Parathenon Project – aims to create a virtual version of the Parthenon and has .... separate elements so that they could be virtually assembled.
- The Ancient Olympic Games – includes three mini applications. First one is the Olympic Pottery Puzzle where the user must reassemble a number of ancient vases by putting together numerous pot shards. Feidias Workshop offers an interactive virtual experience of the construction of a tall golden ivory statue of Zeus. 'The Walk through in Ancient Olympia' is a mini game where the user can make a virtual visit to the site about the games and the ancient building.
- The MuseUs – operates in museums and runs as a smartphone application.
- The Fort Ross – main aim is to archive, distribute, and educate in park history, in an interactive manner for ... ... audience such as park visitors and school students (Georgopoulos et al. 2017, pp. 99–100).

By using the example of an archeological site, Georgopoulos et al. demonstrate how a serious game can be developed with the use of interconnected tools that can disseminate information and acquaint the target markets with heritage settings such as monuments. The ultimate goal is to offer an opportunity for a virtual experience that has objectively authentic expressions. The authors share three applications:

- The first one is about the development of a serious game for the ancient Agora of Athens, which aims to convey the trivial information about the site, while at the same time familiarizes the visitor with the environment and the monuments.

- The second application is about the development of a virtual museum for the Stoa of Attalos (covered walkway in the Agora) that offers the possibility to learn about each exhibit while closely examining them.
- The third application is again about a serious game for the Stoa of Attalos, aiming at the dissemination of conservation options for the most advanced visitors. The game is structured in four parts, which include the introduction, the main game in 3D space, a transitionary section, and a second quiz. A title menu and a closing credits scene are also offered (Georgopoulos et al. 2017, pp. 102–105).

Furthermore, robotics and artificial intelligence are expected to reach their 'tipping points' in the next decade. Digitalization and automation are already visible in the tourism sector. Robots are making their appearance in practically all levels of holiday distribution chains. The primary drivers of their adoption are: productivity, accessibility, and service augmentation. A recent survey examining tourist level of acceptance reports that the majority of international travelers are comfortable with the adoption of robots to facilitate/enrich their travel experience (Murphy, Hofacker, & Gretzel 2017). The key perceived advantages are touted as: the robots' ability to handle data, manage many languages, and function non-stop. Main disadvantages include less job opportunities for humans and the non-personal service of robots.

The international survey also highlights differences in terms of adoption-readiness, with Chinese (92%) and Brazilian (93%) respondents being the most comfortable; while German (37%) and French (47%) respondents were the least responsive. Numerous applications of Robotic technology include service amplification such as chatbots and robotic assistance devices. Chatbots are permeating online communication between customers and vacation intermediaries, aiding users to intermingle with digital assistants, tapping natural language to respond to travel-related queries, and assist in processing bookings (Sheffield 2016). Travel-Bots can be delineated into three categories:

- Customer-service Bots – these are generally integrated into the provider's website and their purpose is restricted to addressing fundamental inquiries and aiding the user to traverse through the homepage (for instance, Sofia-TAP Portugal Airlines);
- Facebook Chatbots – these have a comparatively more interactive function and permit the option to use an interface to enter search and booking-related data (for instance, Expedia's Facebook Messenger Bots); and
- Travel AI (Artificial Intelligence) Bots – these applications are reliant on instantaneous messaging to connect with the customer and they also use algorithms and allow access to information for recommendations (for instance, a virtual travel agent using calendar and email information

> to make tailored recommendations). Robotic Assistance Devices are currently being piloted in travel agencies, Hotels (e.g. Marriott's Butler performing the dual duty of a butler and as an attraction) and Airports (KLM's Spencer – a passenger assistance robot, SITA's self-thrusting baggage robot – Leo). They perform a broad spectrum of functions ranging from networking and entertaining travelers to physically assisting them (such as welcoming, transporting, and checking-in luggage) and guiding them to departure gates (Alexis 2017).

Furthermore, accessibility functions of telepresence technology include the benefits of 3D-holograms and telepresence devices to abolish geographical distance enabling travelers to be virtually existent at any time at any location. For instance, "telepresence devices (e.g. BEAM Pro), Unmanned Aerial Vehicles (UAV) combine teleconferencing, and mobility-technology to create personal 'avatars', allowing travelers to interactively engage themselves while being physically at another location. The main aim and benefit of these technologies is not to replace the actual holiday experience, but to provide increased accessibility. Apart from the obvious application in accessible tourism, other examples involve the possibility to televisit an attraction during night-time/afternoon hours (Alexis 2017, p. 213).

Industrial robots have been popular in the entertainment industry for a long time. For instance, the Royal Caribbean Cruises launched robotic bartenders and they can mix two drinks per custom orders. Another notable example is from the Henn Na Hotel in Japan (Alexis 2017). Initially, the hotel made devoted efforts by underlining a commitment to introducing state-of- the-art technologies to maximize efficiency and pleasurable experiences for its guest: "amongst other innovations (e.g. voice- and face recognition), this hotel was mainly staffed by robots. The reception was staffed by three multi-lingual robots (one of them is a talking dinosaur) responsible for greeting, checking-in and assisting guests. At the cloakroom, a robotic arm stores luggage, and porter robots carry them to the rooms" (Alexis 2017, p. 211). However, the hotel reduced its robotic workforce due to disruptions:

> Japan's Henn na Hotel, which first opened in 2015 with a staff of robots, has cut its robotic workforce after the experience failed to reduce costs or workload for its employees. The hotel, which is located in Nagasaki, will reduce its 243-robotic workforce by more than half and return to more traditional human-provided services for guests, though it will maintain a number of robots in areas where it found them to be effective and efficient. Its change of direction can offer lessons for companies that are pursuing robotic solutions for customer-service roles, reports the Business Insider.
>
> The "firing" comes after complaints from both, the staff and customers. Apparently, a large percentage of the robots were more adept at increasing work for their human counterparts rather than reducing it (Henna na Hotel n.d.).

Digital challenges for cultural heritage institutions

As mentioned earlier, with the development of photogrammetry and computer visualization technologies, digitalization of cultural heritage has advanced from 2D and 3D to 4D. Mechanical digitalization still forms the basis of digital cultural heritage and comprises of digital documentation, research management, and display/visualization and interpretation. Many libraries, archives, and museums hold responsibility to preserve cultural heritage in a given town or city. Despite the much touted opportunities, digitalization has brought with it a unique set of challenges for numerous heritage tourism agencies. For instance, small institutions struggle with basic expenses and tools to successfully launch online presentations and share resources digitally due to lack of human resources, funding, knowledge and training, public sector support, and limited technical infrastructure. From the authenticity and authentication standpoint, technology offers tremendous potential in terms of innovative knowledge sharing platforms, social networks, and artificial intelligence to educate and immerse people into the 'otherness' or 'realness' of heritage. That said, the digital authentication is a complicated process. Numerous challenges associated with digital cultural heritage exist.

According to He, Ma, and Zhang (2017), the digital products often follow a myopic design strategy and can generate non-'user-friendly' experiences. Objects are often taken out of context and showcased only as isolated pieces without any regard for continuity and timelessness of information, for instance the paragon marbles in Greece. It is important to remember that the protection of digital content should not interfere with or serve as a substitute for the protection of physical monuments and heritage sites. Instead, a novel way for public dissemination should be facilitated and used for the commercialization of cultural resources. A more adequate function of digital heritage is the use of modern technologies to develop an open dais for dialogue, exchange, and cooperation between multidisciplinary fields such as archeology, history, art, data information programming, and engineering. Rich cultural and high-tech products can offer deep emotional experiences and high-value services to users/consumers and transport historical heritage back into contemporary life while retaining the inheritance of cultural heritage. All digitization does not necessarily hold heritage value. Simple replication with poorly research data is not authentic in terms of its cultural heritage value. Furthermore, He et al. write:

- Digital heritage should abide by accurate educational benchmarks instead of solely serving as an attractive visual and should represent actual academic narratives and appraised data;
- The goal of digital heritage should be to evoke feelings in the audiences through most advanced scientific and technological tools and to offer products that encompass a high degree of cultural and artistic taste, technology, emotion, and value;

- The use of technology needs to be aligned with research, heritage protection principles, and interpretation and management strategies;
- The role of cultural heritage does not end at the protection of the physical remains of cultural heritage. Rather, regardless of their (degraded) condition, these monuments and sites are full of vigor and vitality that is of practical benefit to the modern customer/user/spectator. The past can serve the present if cultural heritage takes the path towards innovation and national and regional specialization i.e. a new and distinct using innovative and creative global transformation tools in the context of globalization. With the rapid advancement of digital communication and internet technologies, the display of and interaction with cultural heritage is becoming wireless through a cloud-based system – for example the iCloud storage and computing service allow its users to store data on remote servers and share and send to other users (2017, pp. 338–341).

He et al. (2017) call for synergies between cultural heritage institutions and enterprises with expertize in digital technology. Digital heritage needs to be location specific, to make it more authentic, and tailored to each project. Partnering with leading institutions (both, local and global) and traversing transdisciplinary boundaries and a broad spectrum of stakeholders can open the path for sustainable authentication of digital heritage communication systems. Teresa and Marzia provide numerous recommendations to address some of the challenges presented by He et al.:

- *Crowdsourcing and user-based resources* – to facilitate visitor involvement and compatibility with the working staff;
- *Interoperability* – to help reuse digital content and to develop global markets and systems, including those related to tourism, while preventing the undesired effects of Fragmentation;
- *Intellectual property rights – open data, linked data and reuse* – each digital cultural item should be tagged with a license or statement to inform users if and under what guidelines and for which purpose can an object be used (such as commercial, non-profit, personal or academic/research function);
- *Web communication*-social media – identity of the website owner must be clearly stated on the signature page and all other web pages. Multilingual words should be used and the information should be clearly presented and easily accessible;
- *Multimedia and transmediality* – need to enhance traditional information and educational tools with innovative products build from advanced technologies. The purpose is to offer more access and make heritage interactive and engaging both online and via mobile devices or through "multimedia installations" accessible to the public at large;

- *Usability and accessibility* – biggest challenge to e-inclusion is the need for digital qualifications to develop updated strategic technological solutions to improve demand;
- *Synergies and cooperation* – collaboration and co-creation are the key to success in the contemporary era. Identifying key stakeholders and building synergies at local, national, and international levels are arduous tasks (2017, p. 60).

Authenticity of digital heritage

In the aforementioned paragraphs, I have presented examples of numerous applications that support or hold potential to augment digitalization of heritage. I have also pointed out some key challenges from the authenticity and authentication standpoint. This chapter will now examine how the authenticity of digital heritage is negotiated and authenticated. Several authenticating initiatives are shared in recent literature such as:

- Selecting stories and objects that showcase cultural values;
- Facilitating the inclusion of voices from a broader spectrum of stakeholders who have a real connection to the heritage being showcased;
- Offering dialogical platforms and training on how to digitalize oral heritage; and
- Offering an interactive participatory dialogue to elicit views on showcased content.

Innovative and multidimensional techniques are required to conserve and nurture inherited living cultural heritage in optimized conditions. As a case in point, I am sharing the key highlights of the MCHRAP (Makgabeng Community Rock Art Heritage) project in South Africa:

> cultural values of the community are incorporated and ways to mitigate power politics are identified, by engaging numerous stakeholder meetings at all levels of planning and implementation of initiatives. The process of oral heritage first allowed the community to articulate their traditional conceptions of what is valued about the past, what is worth protecting and why. Second, it enabled community members who ordinarily may not have been involved in the protection of heritage to participate. Oral heritage interviews were sometimes conducted in the presence of younger members of the community, a process that encouraged intergenerational dialogue on culture and enabled the transfer of cultural knowledge (Namono 2018, p. 282).

Similar examples can be identified of the manner in which digital preservation is undertaken by cultural organizations such as the World Heritage Center, CyArk, and the Center for Digital Documentation and

Visualization (Owda, Balsa-Barreiro, & Fritsch 2018). The 2D sketches are commonly used to authenticate, showcase, and document patrimonial heritage. A more recent development is the launch of 3D models to enshrine its objectively authentic/historical value by transforming patrimonial data into a visual delight. Owda et al. (2018) illustrate the use of laser scanning and photogrammetry to reconstruct/authenticate photorealistic 3D models. From a sustainability standpoint, "the gradual development of virtual reality, digital photography and three-dimensional information acquisition technology, and the wide application of the Internet in big data, both provide a strong material guarantee for the intangible cultural heritage protection" (Namono 2018, p. 296). The author presents a stepwise path for the digital authentication and protection of intangible cultural heritage (ICH) in a Chinese town:

1 *Improve the intangible cultural heritage digital database, and build a specific rating system* – In the process of the protection of intangible cultural heritage, effective field investigation and data mining system research are carried out, and historical and cultural connotation and artistic characteristics are explored. Moreover, a digital camera, 3D model library, digital platform, mobile terminal, and other digital technologies are applied for the complete record of its artistic form and edition;
2 *Construct digital intangible cultural heritage protection platform* – The media should strengthen the interaction and promote the spread content and quality through micro-blog, WeChat, and other media platforms; and
3 *Establish the intangible museum and protection base* – The digital platform uses 3D model storage and retrieval technology. For example, Fuzhou shadow shows rich humanistic atmosphere in the scenic spot, sets up a live version of the performance and virtual version of digital interpretation and exhibition area and synergizes digitalization and traditional culture to offer enjoyable experiences (Namono 2018, p. 270, 279–282).

In summary, both strengths and weaknesses can be noted in the collaborative digitalization of heritage. Strengths include: (a) benefit, from cooperative solutions, for those without appropriate IT infrastructure; (b) provision of a digital base, besides space for digital presentation, for data exchange with big platforms; (c) benefits for those who offer cooperative infrastructure in terms of profit, volume, and relevance of the entire collection; (d) smaller institutions are able to digitalize; (e) smaller institutions are able to focus on content as bigger institutions offer central planning and management and free them from administrative duties; and (f) statewide coordination of projects and transfer of knowledge (Preuss 2016). Successful examples can inspire and propel more institutions to follow suit, enabling a broader sustainable level of cultural/heritage authentication. Nevertheless, several weaknesses can be evidenced on the supply side of the heritage equation such as:

- Principle of subsidiarity within the cultural sector – every government (federal, state, or municipality) is responsible for the development and funding of culture in its jurisdiction. The outcome is that the institutions are confined within the budgetary restraints of the local municipality;
- The projects are not viable in the long term if there is no financial commitment from the state or access to third party funds; and
- It is challenging to secure cooperation and communication between different institutions (Preuss 2016).

From positive and negative standpoints, few studies have examined the impact of digitalization of heritage on authenticity. Clearly, negotiated authenticity can become the mediating point as heritage transcends to another platform where it can be showcased remotely from a distance.

Prospects for sustainability of digital cultural heritage

For any innovative initiative to generate a lasting and meaningful benefit, sustainable mechanisms and a strategic game-plan are required. Many projects get shelved because of budgetary constraints. Funding is crucial, in an ongoing manner, and this is a big challenge for nonprofit cultural institutions across the globe. In fact, financial crisis can impede progress and success of digital initiatives. Long-term commitment of universities and cultural institutions is also paramount. An interdisciplinary approach can help garner the much needed political and public support for the cultural/heritage community; rather than secluded initiatives, a collective/collaborative scheme is beneficial for the success and development of sustainable solutions. A collaborative approach can serve as a conduit between human resources, expertise, and IT infrastructure by cooperative and interdisciplinary think tanks and ideologies/actions. At the same time, assimilation of information from various fields and establishments requires a common set of strict guidelines, devotion, and inter-sectoral collaboration (Preuss 2016).

Particularly, during the digitalization process, content of objectively authentic heritage projects needs to be appropriated/negotiated in a sanitized and sustainable manner. The aim is to accomplish strategic and optimal tradeoffs that can enshrine the core essence of actuality (genuineness). As contended by Calcagno and Zabatino, "in the context of sustainability, the process of co-production becomes more complex and has to be managed" diligently to facilitate creativity and innovation, while ensuring the empowerment of the local artists/producers (2010, p. 2). The authors stress on different levels of sustainability such as 'bottom-up' and 'open production' to authenticate collaborative digital heritage in a negotiated manner. The third level of sustainability, a processual form, is recommended that can preserve artistic control over the process while facilitating learning and participant involvement/engagement.

As we have seen, a collaborative approach is one of the essential pillars of heritage sustainability (Chhabra 2010) and it can aid in optimizing a meaningful tradeoff between preservation and economic remunerations. ICT can play a crucial role as a stimulating mediating factor. In the heritage sector, especially in developed countries, advanced adoption of ICT is evident albeit in developing countries, online presence of most heritage attractions is extended to a mere website. Many popular heritage tourism places in developing countries lag behind in embrace of advanced ICT. Most of tangible and intangible heritage on sale is produced by small and micro enterprises (SMEs) and these often lag behind in digital innovations and require ongoing public sector support to sustain income. Several push and pull factors have shaped the implementation of IT by SMEs:

> Push factors are related to the risks and threats of not using technological enabled products and solutions, whereas the pull factors are allied to the benefits that can be enjoyed by businesses in implementing technology. Some of the push factors include new government policies, the pressure exercised on businesses by global marketplaces, and new education and training of employees in the tourism sector. On the other hand, some of the key pull factors offer benefits from increased connectivity and opportunities to synergize with customers and businesses. (Konstantinou 2016, p. 1559)

Cost, risk, lack digital expertise, and digital exclusion are some of the barriers to the adoption of technology identified by recent literature (Chhabra 2015; Díaz-Andreu 2017; Konstantinou 2016). Díaz-Andreu (2017) lament that the digital media is facilitating new ways of exclusion because it is impossible for everyone to develop technological expertise, and remote communities do not have access to the world wide web (www). However, from a positive viewpoint, Konstantinou also writes that:

> technology and infrastructure have transformed the traveling experience to a seamless or 'door' to 'door' travel. Technology is leading to the emergence of new business models such as shared economy, or else people-to-people and 'collaborative consumption,' which usually refers to a new socio-economic system of relationships among people that harness the unique power of technology. New online platforms are used to connect people and marketplaces through new forms of transactions such as home sharing and these business models and transactions have significant impact on the environment and the society. (2016, p. 1558)

Furthermore, according to Ukpabi and Karjaluoto, "the ubiquity, flexibility, personalization and dissemination features of mobile technology make it a veritable tool for both marketers and consumers in tourism and hospitality services (Kim, Park, & Morrison 2008). For consumers, the functionality of mobile technology, such as the ease of access to travel information and trip guides, is an essential feature of its increased adoption, whereas for marketers, it is the opportunity to send marketing messages to a targeted audience. However, individual differences shape mobile technology adoption in tourism and hospitality services. For instance, Kim et al. (2008) report that experienced and frequent travelers have a higher rate of mobile technology adoption than inexperienced infrequent travelers" (2017, p. 620). More research is required to examine the potential and threat of the aforementioned technological advancements in their capability to rationally authenticate heritage tourism repositories.

Furthermore, recent literature reports surging demand for cultural expression in heritage tourism experiences, especially in the form of ICH. Chhabra postulates that ICH is:

> a distinct selling asset for host communities and a key generator of satisfactory heritage tourism experiences for tourists. It is argued that ICH is deeply guided by the objective version of cultural authenticity and can … … make a heritage destination competitive … … However, it needs to be regarded and developed as a sustainable tourism commodity, … … … … … and ongoing initiatives are needed to protect the 'cultural continuity of communities'. (2019, p. 2)

Dimitropoulos et al. (2018) offer an innovative holistic agenda for protecting and broadcasting ICH beyond the opportunities rendered by digital platforms. The terms 'Living Human Treasures' coined by UNESCO refer to the cultural custodians who have the knowledge and skill to enact or recreate selected parts of ICH. Dimitropoulos et al. explain that "ICH creations are transmitted orally and/or by gestures and usually undergo modifications over a period of time, resulting to a process of collective recreation" (2018, p. 4). Hence, strategic efforts are required to safeguard ICH. Dimitropoulos et al. present a system (i-Treasures) that does more than digitalizing the ICH content. It innovatively informs of novel ways of protecting ICH expressions in domains where human mobility is enormously useful, such as performing arts and handicrafts. The authors delineate and examine numerous ICH manifestations from different European regions such as traditional singing, traditional dances, craftsmanship, and contemporary music composition. Table 4.2 offers an overview of numerous scenarios to illustrate its application:

- The essence of i-Treasures lies in the identification of particular traits (such as postures and audio patterns) in a variety of ICH expressions using multi-sensor technology (for instance, cameras, mocap systems, and microphones);

Table 4.2 Application scenarios of I-Treasures System

In a typical application scenario, such as the case of a traditional dance, an expert can use the system to (a) capture a dance performance and store medium-level features (e.g. dance steps) and high-level concepts (e.g. dance style) to the database of the system and (b) create educational courses to teach the recorded dance.
Particularly, a variety of mocap technologies can be supported, such as Kinect sensors, optical motion capture systems. Next, the recorded motion can be analyzed, through probabilistic inference, to extract numerous of medium-level traits, linked to the motion of the dancer (for instance, in the case of the traditional Tsamiko dance, the "single" or "double" step is distinguished to ensure the recognition of the dance variation or the style).
Furthermore, educational courses can be designed using didactic modules or creating new 3D games to facilitate dance learning and its practice, with the help of a unique using a novel structure that can support the rapid design and development of motion-centered game type of applications.
On the other hand, a learner or researcher can (a) have access to the content of the platform and search for a specific dance in the repository of the system and (b) use the interactive learning applications of the i-Treasure platform. In the former case, the user can retrieve general information about a specific dance (text, images, and videos) as well as its medium-level features and high-level concepts.
Moreover, the user can search for specific medium-level features (e.g. dance figures) to find possible similarities of the dance with other dances. Each game is designed following a specific learning scenario, defined by an expert, and supports a mechanism for the online evaluation of the player's performance.

Source: Dimitropoulos et al. (2014)

- Next, data fusion analysis is used to utilize information from the different modalities;
- Furthermore, content and context are integrated to transform the mined data into interpretation levels that can be comprehended by humans;
- This information, paired with other cultural resources, is accessible from the i-Treasures Web platform, which offers a variety of content (for instance, text, audio, images, and video) with different supportive roles: expert, learner, and researcher; and
- Furthermore, sensorimotor learning offers a learning platform to aid training and evaluation (Dimitropoulos et al. 2018).

Closing comments

Review of recent literature reveals that every country differs in the extent and manner to which its heritage institutions have the capacity and resources to embrace ICT. For instance, many heritage institutions in Europe lag behind and have yet to extensively embrace online e-commerce applications. Most heritage institutions in the United States, on the other hand, continue to make advanced use of ICT in the form of AR and VR and mobile apps. Japan is ahead in technology whereas China has been at the strategizing and planning stage. Russia, on the other hand, remains far behind. Asian countries such as

India, Nepal, and Pakistan are constrained by limited resources and budgetary restrictions.

Literature offers several examples of how digitalization can promote sustainable use of cultural and heritage resources. In fact, most recent literature from the pandemic time shows overwhelming focus on equitable showcasing of past histories to engage a broad spectrum of the audience. One example is the virtual 'Whole Story', pay-what-you-wish program by the Phoenix Art Museum (Arizona, USA): This live-stream storytelling show celebrates the nuances of our humanity through personal stories told from the Black perspective (phxart.org 2020). Numerous scenarios of harmonious digitalization and sustainability initiatives, associated with fragile cultural/heritage resources, can be noted. Using the example of the Rock Art Museum, Namono writes:

> Using digital technologies in the process of collecting and documenting oral heritage allows previously marginalised voices to feed into heritage and historical narratives. Literary heritage narratives have tended to dominate the dissemination of information on African heritage, whereas African cosmologies and oral traditions are the intangible values of place that attract visitors to heritage sites. In the Makgabeng, oral heritage narrated through stories, songs, dances and poetry and collected using digital technologies has helped preserve African values threatened by the onslaught of Western ones, especially through written European languages and social media. The Makgabeng Community Rock Art Project has re-valued the role of elders in sustainability of heritage tourism initiatives and the integration of a community structure as a sustainable ready-made framework to heritage management in Africa (2018, p. 269).

Although reverse trend is noted in the wake of the ongoing pandemic, it is important to discuss the discourse on digital detoxification in the last years of the pre-pandemic era. Alongside stories of fascinating digital innovations, it is impossible to overlook parallel voices that were calling for digital detoxification of heritage tourism experiences. There has been an emerging concern over the manner ICT is transforming the cultural landscapes by dehumanizing heritage tourism experiences and producing couch surfers and digital nomads. In the context of heritage tourism, interrogations have included: can technology adequately replace the 'human-face' of the authentic tangible and intangible experiences of heritage tourism? To what extent are heritage tourists willing to interact with robots in their desire for interactive engagement with heritage resources? What is the perceived optimal trade-off between dehumanized efficiency and authentic experience through human interactions? What kind of negotiated authentication process can optimize its use in human settings? As we approached 2020, the development and diffusion of robotics and artificial intelligence in heritage tourism experiences services ignited further deliberations associated with the sustainable function of ICT in authenticating and humanizing heritage tourism.

However, today as the pandemic looms over the entire globe, technology has become a blessing in disguise, enabling virtual connections with friends and relatives, the immediate community, workplace, news, and the world. It has also made possible for destination marketing organizations and cultural/heritage institutions to deliver heritage at home, in terms of knowledge sharing and engaging/immersive virtual experiences. The cultural and heritage repositories, museums, and other heritage institutions are unpacking and opening their treasures for digital viewing. It is early to perceive the extent to which authenticity will be negotiated to relate to the current environment and make heritage meaningful in the face of current political, social, and economic disturbances.

There is a parallel call for retrospection, a time of reckoning and shaming of hypocritical promotions of sustainability that have compromised local communities, wildlife, and the world heritage. Some scholars find the pandemic an opportunity to pause, take a backward step, and develop a reformative agenda to put heritage tourism on an authentic path. In this regard, a notable insight is offered by Evans (2020): As millions of travel and tourism workers now find themselves out of jobs, furloughed, or, as in the case of some cruise ship employees, stuck indefinitely at sea, the industries involved seem to be gearing up for some future "recovery", insinuating a return to the baseline of pre-coronavirus. That simply cannot happen because the pre-coronavirus travel and tourism industries will not function in a post-coronavirus world. Everything must change: the way we fly, the way we dine, how we wait in line – even how we go to the beach. Our very concept of vacation may have to change.

References

Alexis, P. (2017). R-tourism: Introducing the potential impact of robotics and service automation in tourism. *Ovidius University Annals, Series Economic Sciences*, *17*(1), 211–216.

Anderson, E. F., McLoughlin, L., Liarokapis, F., Peters, C., Petridis, P., & De Freitas, S. (2010). Developing serious games for cultural heritage: A state-of-the-art review. *Virtual reality*, *14*(4), 255–275.

Arsul, A., Lumenta, A., & Sugiarso, A. (2015). E-tourism Kabupaten Pulau Morotai. *Jurnal Teknik Elektro dan Komputer*, *4*(3), 71–78.

Boo, H. (2020). A Digital Future for Culture and Heritage. Arts Management and Cultural Laboratory: Retrieved from https://amt-lab.org/blog/2020/3/a-digital-future-for-cultural-heritage.

Calcagno, M., & Zabatino, A. (2010, October). Open innovation and sustainable value in cultural productions. In *ESA Research Network Sociology of Culture Midterm Conference: Culture and the Making of Worlds*.

Chhabra, D. (2010). *Sustainable marketing of cultural and heritage tourism*. London: Routledge.

Chhabra, D. (2015). *Strategic marketing in hospitality and tourism: Building a'SMART'online agenda*. UK: Nova Science Publishers, Inc.
Chow, I., & Murphy, P. (2011). Predicting intended and actual travel behaviors: An examination of Chinese outbound tourists to Australia. *Journal of Travel & Tourism Marketing*, *28*(3), 318–330.
Cosmas, J., Itegaki, T., Green, D., Grabczewski, E., Weimer, F., Van Gool, L., ... Kampel, M. (2001, November). 3D MURALE: A multimedia system for archaeology. In *Proceedings of the 2001 conference on Virtual reality, archeology, and cultural heritage* (pp. 297–306).
Díaz-Andreu, M. (2017). Introduction to the themed section 'digital heritage and the public'. *International Journal of Heritage Studies*, 23, 404–407.
Dimitropoulos, K., Tsalakanidou, F., Nikolopoulos, S., Kompatsiaris, I., Grammalidis, N., Manitsaris, S., & Hadjileontiadis, L. (2018). A multimodal approach for the safeguarding and transmission of intangible cultural heritage: The case of i-treasures. *IEEE Intelligent Systems*, *33*(6), 3–16.
Evans, A. (2020). Coronavirus will change how we travel. Retrieved on July 25, 2020 From: https://www.nbcnews.com/think/opinion/coronavirus-will-change-how-we-travel-will-probably-be-good-ncna1186681
Georgopoulos, A., Kontogianni, G., Koutsaftis, C., & Skamantzari, M. (2017). Serious games at the service of cultural heritage and tourism. In *Tourism, culture and heritage in a smart economy* (pp. 3–17). Cham: Springer.
Gervautz, M., & Schmalstieg, D. (2012). Anywhere interfaces using handheld augmented reality. *Computer*, *45*(7), 26–31.
He, Y., Ma, Y. H., & Zhang, X. R. (2017). "Digital heritage" theory and innovative practice. *International Archives of the Photogrammetry, Remote Sensing & Spatial Information Sciences*, *42*.
Jung, T. H., & tom Dieck, M. C. (2017). Augmented reality, virtual reality and 3D printing for the co-creation of value for the visitor experience at cultural heritage places. *Journal of Place Management and Development*, *10*(2), 140–151.
Kim, D., Park, J., & Morrison, M. (2008). A model of traveller acceptance of mobile technology. *International Journal of Tourism Research*, *10*, 393–407.
Konstantinou, J. K. (2016). Digitization of European SMEs in tourism and hospitality: The case of Greek hoteliers. *World Academy of Science, Engineering and Technology, International Journal of Social, Behavioral, Educational, Economic, Business and Industrial Engineering*, *10*(5), 1558–1562.
Kotler, P., Bowen, J. T., Makens, J., & Baloglu, S. (2017). *Marketing for hospitality and tourism* (Global ed., 7th ed.). Pearson.
Liarokapis, F., Petridis, P., Andrews, D., & de Freitas, S. (2017). Multimodal serious games technologies for cultural heritage. In *Mixed reality and gamification for cultural heritage* (pp. 371–392). Cham: Springer.
Murphy, J., Hofacker, C., & Gretzel, U. (2017). Dawning of the age of robots in hospitality and tourism: Challenges for teaching and research. European Journal of Tourism Research, 15(2017), 104–111.
Namono, C. (2018). Digital technology and a community framework for heritage rock art tourism, Makgabeng Plateau, South Africa. *African Archaeological Review*, 1–16.
Natale, M. T., & Piccininno, M. (2018). Italy: Tourism and technological innovation: The spectacularization of cultural heritage in Rome and Cerveteri. *Uncommon Culture*, *7*(1/2), 134–145.

Owda, A., Balsa-Barreiro, J., & Fritsch, D. (2018). Methodology for digital preservation of the cultural and patrimonial heritage: Generation of a 3D model of the Church St. Peter and Paul (Calw, Germany) by using laser scanning and digital photogrammetry. *Sensor Review*, *38*(3), 282–288.

Pletinckx, D., Callebaut, D., Killebrew, A. E., & Silberman, N. A. (2000). Virtual-reality heritage presentation at Ename. *IEEE MultiMedia*, *7*(2), 45–48.

Popescu, M. A., Nicolae, F. V., & Pavel, M. I. (2015). Tourism and hospitality industry in the digital era: General overview. In Proceedings of the International Management Conference, *9*(1), 163–168. Bucharest, Romania: Faculty of Management, Academy of Economic Studies.

Preuss, U. (2016). Sustainable digitalization of cultural heritage – Report on initiatives and projects in Brandenburg, Germany. *Sustainability*,*8*(9), 891.

Rullani, E. (2010). *Modernità sostenibile*. Marsilio, Venezia.

Sheffield, J. (2016). The Ultimate Travel Bot List. 30 Seconds to Fly Homepage. Retrieved on May 30, 2017 from: https://www.30secondstofly.com/ai-software/ultimate-travel-bot-list

Ukpabi, D., & Karjaluoto, H. (2017). Consumers' acceptance of information and communications technology in tourism: A review. *Telematics and Informatics*, *34*(5), 618–644.

5 Smart and sustainable marketing frameworks in heritage tourism

This chapter scrutinizes sustainable marketing blueprints in heritage tourism and offers insights on how authenticity features in the existing marketing agenda. Furthermore, authentication of brand authenticity is covered and implications of COVID-19 on sustainable marketing of heritage tourism are discussed.

Authenticity has become a trendy marketing catchword today. It is not only employed to market places and environments but heritage merchandize as well (Hede & Thyne 2007; Kolar & Zabkar 2010; Napoli, Dickinson, Beverland, & Farrelly 2014; Timothy 2011). Restaurants, gift shops, art vendors, and heritage hotels, all use authenticity as a marketing ploy to attract their clients. According to Timothy, "heritage and tourism managers have learned that authenticity sells and genuineness is good business" (2011, p. 290). There is a need to strategically plan marketing planning for sustainable outcomes so that places and products can be promoted in a manner that the historical or structural veracity of heritage sites and commodities is not compromised. Hargrove (2003) writes that while debate looms about how to appraise authenticity, it is important to be mindful of the fact that a customer expects that the heritage sites will be preserved and showcased with integrity and truth. Political mandates should not shadow efforts to develop authentic experiences for customers.

Many countries across the globe are using cultural heritage as a platform for urban renewal, economic viability, and for forging a shared sense of identity among host communities (Chhabra & Zhao 2015; Donohoe 2012; Timothy 2011; Timothy & Boyd 2003). However, heritage for the purpose of tourism can be a double-edged sword (Butler & Inanovic 2016). On one hand, exponential demand has provided political and economic base for securing government and fiscal support to expand conservation-friendly activities. On the other hand, issues such as surge in visitation, overuse, or inappropriate use of cultural assets have threatened the survival and authenticity of fragile assets (Chhabra 2010). As an instance, a vicious circle of heritage tourism development is noted in Venice (Italy), where surge in the number of day visitors has encroached on the quality experiences of the 'high-end cultural tourist' market. Dual marketing strategies have been suggested: demarketing of day trippers and niche marketing to attract the high-end visitors.

Strategic management of heritage resources has become a crucial objective in the sustainable development initiatives (McCamley & Gilmore 2018; Timothy 2011; Timothy & Boyd 2003). Issues of concern include restricted development, management of disputed land use of heritage resources, equitable access to heritage resources, and responsible use of infrastructure and other facilities within the heritage tourism support system (Coccossis 2008). Numerous remedies have been suggested such as carrying capacity measures to manage and control visitor numbers (Hunter 1997), appropriate training and educational programs (Timothy 2011), and active civic engagement (Chhabra 2010). Cultural/heritage management is, therefore, an ongoing/evolutionary process that requires consistent monitoring (Jamieson 2000). Jamieson (2000) emphasizes on the significance of conservation, accurate interpretation techniques, objectively authentic experiences, and viable economic health. For the heritage industry in Canada, Boyd identifies five areas of focus: "mutually beneficial partnership; a national strategy with local linkage; integration of private and public sectors; knowledge-based communities; and greater attention to culture and heritage within the context of a wider view of tourism in general" (2002, p. 228). Clearly, sustainable marketing of heritage tourism needs to integrate multiple criteria into its agenda that promote conservation, mutually beneficial partnerships, holistic view, local connections, effective interpretation, and economic benefits for the host community, knowledge management, and authenticity.

Key highlights of this chapter are:

- Sustainable marketing blueprints in heritage tourism
- Authenticity in the marketing agenda
- Authentication of brand authenticity

Wide-ranging viewpoints exist on sustainable development of heritage tourism which propels attention toward sociocultural, political, and economic issues (Mowforth & Munt 2016). Despite lack of homogeneity, some common themes are symptomatic across the spectrum of heritage tourism that illuminates the need for a harmonious path between conservation and commodification. Hunter (1997) places emphasis on achieving sustainable goals that resonate with the economic progress of host communities but minimize negative impacts. Most sustainability objectives are underpinned on the principles and guidelines set forth by Agenda 21 and the emerging themes include:

- Longer viability and quality of natural and human resources;
- Reducing friction in the complex interactions between stakeholders of diverse interests;
- Adhering to host population needs and quality of life;
- Benefiting the future generations;

- Balancing visitor numbers based on preservation guidelines to maintain long-term biological and cultural diversity;
- Reassessing role of tourism in host society; and
- Maintaining cultural integrity (Chhabra 2010, p. 21).

Marketing strategies, inclusive of a sustainability agenda, can inspire responsible consumption of heritage resources and help to inculcate moral values and mindfulness among the visitors, toward the host community, and the visited environment (Chhabra 2010; Donohoe 2012; Font & McCabe 2017). The American Marketing Association widely advocates the inclusion of societal obligation in its definition of marketing: "as an activity, set of institutions, and processes for creating, communicating, delivering, and exchanging offerings that have value for customers, clients, partners, and society at large" (AMAssociation 2007). Moreover, the influence of macro environment factors on heritage tourism policies and marketing agenda cannot be denied. Parson and Maclaran (2009) write that the macro environment is prominently shaping consumer identity in the contemporary era and its key defining elements are: (1) fragmentation (dissolution of mass-market stances resulting in detachment and numerous market segments), (2) de-differentiation (failure of conventionally ordered hierarchies and clouded boundaries), (3) hyperreality (demand for fake experiences), (4) chronology (seeking solace from past), (5) pastiche (seeking hybrid lifestyles), (6) anti-foundationalism (the conformist views are being questioned and green/sustainable products and experiences are being demanded), and (7) pluralism (demand for diverse and multicultural environments). However, as COVID-19 continues to loom over our lives, some trends might fade away such as hyperreality, pastiche, and pluralism. New trends (such as popularity of social robots while at the same time personalizing marketing, digital conferences, virtual immersions, surge in domestic travel because of strict international travel regulations, social networking, preferences for home delivery and staycations in the home yard, meeting people digitally in the comfort and safety of their homes, and co-creative synergies with other players in the supply chain) are emerging, and time will tell if these are fads or will fortify novel changes, as vaccine develops and travel transforms to a new normal. Several scholars note that COVID-19 ramifications will be far reaching and reformation will be key for the successful launch of post-pandemic heritage travel. Undeniably, the COVID-19 has changed everything about marketing and an innovative fluid "mindset (rolling iterations based on quick-paced and changing marketing conditions), premised on sympathy, empathy, respect and trust", is required of cultural and heritage institutions (Phelan 2020).

Yet, heritage tourism institutions have, for several decades, suffered from marketing myopia as they view marketing as a stigma slanted toward profit-centered objectives (Chhabra 2015; Misiura 2006). To date, most heritage tourism institutions have been slow to embrace a holistic approach toward sustainable marketing. Recurrent themes, beyond the last decade, in heritage tourism marketing literature were mostly short sighted (Donohoe & Needham 2008) and limited to: market segmentation (based on sociodemographic characteristics, spending propensity, motivation, activities, attitudes/perceptions, and benefits), consumer behavior, communication (internal and external), and promotion (predominantly featuring advertising and effectiveness).

However, in the more recent times, many scholars of heritage tourism have argued that marketing can assist in resolving issues associated with crowding, conservation, authenticity, interpretation, and those concerning the society at large (Chhabra 2015; Font & McCabe 2017; Timothy 2011). Especially, a tourism systems approach toward heritage marketing, underpinned on synergies between different stakeholders of tourism, can lead to innovated and value added initiatives (Chhabra 2015). Donohoe and Needham (2008) argue that responsible marketing is particularly complimentary to the sustainable tourism agenda. The sustainable marketing perspective particularly stretches beyond the conventional threshold and embraces a holistic perspective associated with economic, social, and environmental responsibility. In the heritage tourism context, this triple bottom line stance aims of establishing a tradeoff between economic numerations/visitor satisfaction and responsible use of local resources. The goal is to identify/develop alternative strategies that discourage exploitation of local heritage and consider the overall well-being of the society. Donohoe and Needham (2008) write that:

> In heritage tourism, the audience goes to the product for experience. Admission is free at many heritage sites and this may or may not create a marketing advantage over competition. Promotion communicates the key features and benefits of the product to consumers. Place represents the location where the product can be purchased and it is commonly referenced in the marketing literature in conjunction with distribution. In the case of heritage sites, consumers are brought to the product through promotions that feature the sites as iconic and unique tourism destinations. This marketing mix has been widely debated in the heritage tourism literature and has generated an awareness of the social marketing concept … built on the classic marketing mix but it deferred short-term consumer satisfaction to long-term consumer welfare and it is suggested that businesses include ecological and social considerations in product development and marketing. The argument is that marketers need to meet changing social and ethical responsibilities in order to protect the producer's long-term interests and to avoid failure (2016, p. 2).

Gradual awareness of outreach potential of marketing by heritage institutions has dulled its stigmatized image and propelled its acceptance as a useful tool for promoting sustainable use of fragile heritage resources. For instance, Misiura's (2006) work centralizes the significance of customer needs and high-quality service. But her model also considers the impact of external (macro) environment factors on heritage institutions, visiting markets, and the marketing mix in general. Misiura outlines the need for identity promotion through inclusive heritage initiatives to effectively appeal to responsible target markets. She uses the term 'nostophobia' to argue that there is a past to be shared so it should be promoted and commercialized and that an organization should strive to build corporate identity drawn from key heritage representations. However, by limiting the parameters to a consumer-centric and corporate-centric discourse, her work disregards the risks associated with commercialization of local heritage resources. For instance, power interplay can undermine the inclusion of local voices. As stated by Rowan and Baram (2004), power dynamics shape marketing in terms of production, distribution, and consumption of heritage. By referring to heritage tourism in consumerist terms, the authors take a global view to scrutinize the relationship between materialistic aspects of past. They add an economic perspective to the tourist demand for the past and archeology and caution that this stance can compromise the original purpose of past and archeology. McLean (2002) discusses heritage marketing in the context of museums and argues that the core purpose of museums, in the context of conservation of objective authenticity, should guide the marketing strategy. In other words, it is amplified that marketing is a strategic tool for survival and revenue generation, and can assist in promoting conservation tactics and forging a strong bond with the public at large. A chronological ideological shift can be noted in the work of the three authors from product-centric marketing drawn from the traditional purpose of museums (McLean 2002) to a cautionary standpoint regarding the influence of power dynamics (Rowan & Baram 2004) in shaping heritage representations to stressing on a consumer-centric (marketing) approach (Misiura 2006) while considering the macro environment factors. Clearly, marketing has stirred a dialogue within heritage institutions as they struggle to embrace it, mostly for economic reasons in the face of budget cuts and dwindling financial support.

More recent work has extended the heritage marketing paradigms further to embrace sustainability themes such as co-creation, stakeholder collaboration, host community partnerships, visitor mindfulness, conservation, and authenticity. For instance, Gilmore, Carson, and Ascencao (2007) discuss sustainable marketing initiatives pursued by a World Heritage Site (WHS). Their first argument is that the WHS is not a conventional product per se but is marked by prominent service characteristics to offer intangible social experiences that are spontaneously produced and consumed. Compelling emphasis is placed on appropriate and effective promotional and interpretive content. Socioeconomic dimensions and environmental perspective are positioned at the opposite ends of the tourism spectrum in that the former aim to facilitate tourism visitations, revenue, and value whereas the latter is focused on protecting the cultural and natural environments. The authors do not alienate between the cultural and environmental dimensions and argue that environmental objectives should not dictate the marketing agenda. Instead, a sustainable tourism service product should be reliant on "interactive management, planning and decision making and the implementation of a consistent service product to large groups of people in a well-managed way" (Gilmore et al. 2007, p. 254).

Broader sustainability goals can assist in aligning priorities of marketing professionals of tourism organizations (such as heritage and other support sector providers, intermediaries, retailers, competitors and trade associations, the government, and the voluntary associations) with those of the conservationists and cultural custodians, to forge strong bonds. Some studies have suggested that strategic management of visitors and their activities can be channeled through informational/promotional materials so that they can be informed and guided toward sustainable behavior (Timothy 2011). Fullerton, McGettingan, and Stephens (2010) conduct a two phased study in Ireland to first examine the linkage between marketing and visitor management and then gather insights on perceived impact of site alterations, routing of visitors, and staged orchestrations. Chhabra (2009) tests a sustainable marketing paradigm by examining the marketing plans of museums from across the United States. She incorporates strategic marketing elements such as environment analysis, local community involvement, partnership, and maintenance of traditional preservation-based objectives. The author reports a gap between the proposed model and existing marketing strategies. Myopic emphasis and current survival are reported in many museums. In a nutshell, Chhabra's study points out that although sustainable meticulousness is still slow to permeate the way heritage tourism functions, it is not impossible; this can be attributed partly to visitor management strategies and partially to the growing heritage audience who is becoming more attentive to environmental and other negative impact-related issues. The heritage institutions should not deviate from their core purpose of conservation and safeguarding of objective authenticity. A long-term approach, for instance, can help museums strategize deployment of their resources in a manner that integrates conservation agenda into the marketing plans. Technology and authenticity and its authentication process have become strategic elements of sustainable marketing paradigms in the contemporary times (Figure 5.1).

A revised strategic sustainable heritage tourism marketing framework is presented by Chhabra (2010) by adding four new elements: authenticity, interpretation, creating mindful visitors, and economic viability. Significance of the role of interpretation in marketing and promoting sustainable use of heritage sites, in terms of effectively managing visitors, is increasingly documented in heritage tourism literature (Moscardo & Woods, 1998; Timothy 2011; Timothy & Boyd 2003). Sustainable interpretation techniques ensure that the core message of an exhibit or story is not compromised. Cogswell writes that content in "brochures, guides, and other interpretive materials should contain good general explanations balanced with concrete local reference or illustrations. Exhaustive display of collections, uncaptioned photographs, and disorganized 'show and tell'

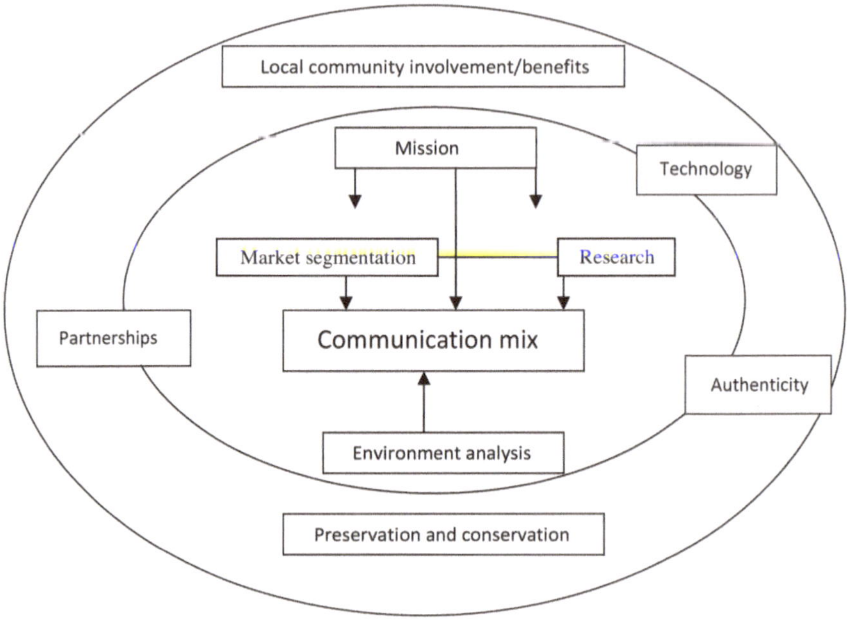

Figure 5.1 Sustainable marketing framework for heritage tourism.

Source: Chhabra (2009, p. 313)

strategies often detract from the key goal of interpretation" (1996, p. 9). To this end, the local manpower can be utilized and examples include cultural custodians, local dialect, and local guides.

Worthy of inclusion also are artists, local craftsmen, folk music, dancers, and heritage commodities such as clothing, decorations, etc. Folk people and cultural artisans need to be strategically cultivated through relationship building efforts. Visitor mindfulness refers to a receptive state and willingness on the part of the visitors to be educated, view, apprehend, and value the past from multiple standpoints (Moscardo 2001). Several techniques exist to encourage visitor mindfulness and should promote: (1) willingness to learn; (2) attentiveness of the setting; (3) ongoing development of new routes and techniques; and (4) development of conditions that are harmonious with the setting, objective authenticity of the site, appealing messages and adaptivity to changing environments (Moscardo 2001). What is required is a holistic plan, employing a multidisciplinary perspective and a diverse group of stakeholders. Chhabra tests her advanced model of sustainable marketing on heritage sites, festivals, souvenirs, and heritage hotels across the globe. Table 5.1 shares progress in sustainable marketing. Ongoing issues of relevance/debate include: politics of heritage; identification of relevant stakeholders; opportunities presented by the emerging middle class across the world; authenticity as an emerging marketing brand;

Table 5.1 Progress in sustainable marketing

• There is universal support for the objective/essentialist version of authenticity of heritage. However, delicately negotiated versions are also deemed important to capture the attention of the contemporary audience. • Global support for the preservation and restoration of heritage continues to exist. • Marketing is still in its initial phase at most heritage institutions in developing countries, such as India, Sri Lanka, and Nepal, although it is actively leveraged as commercial tool in developed countries. A perusal of more recent literature shows that several developing countries are leaning toward consumer-oriented marketing in the near future, whereas the developed countries have now slanted toward integrating societal and tribal marketing techniques with commercial marketing. • Sustainable marketing of heritage is embraced, to some extent, in heritage institutions across the globe. Simultaneously, a need has emerged for appropriate training and support to develop strategic heritage marketing initiatives and plans from the public sector. • Strategic planning, which facilitates civic engagement and long-term economic viability, has become a crucial component in sustainable marketing plans. • Partnerships and inter-sector collaborations between key stakeholders mostly remain ignored, although they are being acknowledged as foundational tools of sustainable marketing.

Source: Chhabra (2010, 2015).

technological opportunities and pitfalls; effectiveness of alternative marketing paradigms such as tribal marketing, demarketing, select marketing, and educational marketing; and ethics of marketing.

Based on observation and outcome of her case studies, Chhabra has incorporated more elements in her revisited strategic sustainable heritage marketing model by including civic engagement, exclusivity, and support for enterprise engagement to promote cultural, social, and economic health of heritage tourism. Several studies have confirmed the significance of different elements suggested in the SSHTM model (Donohoe, Salo, & Gilmore 2016; McCamley & Gilmore 2018). Donohoe et al. (2016) offer a critical assessment of marketing strategies pursued by heritage site (the Rideau Canal World Heritage Site) in Canada. The authors note that from a demand standpoint, the market for sustainable products remains limited in Canada and across the globe. However, the demand for heritage tourism has grown exponentially despite dwindling demand for sustainable merchandize. In other words, the majority of the growth is represented by casual or sightseeing cultural tourists who remain unaware of the heritage value of the visited destination. While the purposeful cultural tourist represents a minority in heritage site visits, their quest for a deep culturally integrated experience is influencing a change on the supply-side. Heritage institutions are leaning toward sustainable management regardless of drop in revenue. They are seeking training and information tools to embrace sustainability.

While sustainability issues remain non-essential to most product marketing managers, a paradigm shift is happening due to academic and practical support for sustainable heritage tourism marketing. The Rideau Canal case study illustrates that both demand and supply side of the heritage tourism market is recognizing the need to embrace sustainability. This offers an optimist window for the Rideau Canal and other heritage sites around the world. However, internalizing and integrating sustainability as a 'standard operating procedure' will require a long-term commitment. As reported below, an important take away point is that successful sustainable marketing initiatives require internalization (embedded in the day-to-day management operations) and long-term investment in terms of time and finances:

- *Demand satisfaction* – Demands for tourism experiences in increasingly sensitive natural and cultural environments are increasing. Marketing transparency is ideal for attracting and satisfying customer demand. Tourism opportunities are aligned with current demand trends (monitoring of satisfaction, demands, etc.);
- *Achieving competitive advantage* – Focusing on delivery of sustainable outcomes and maintenance of heritage integrity and/or heritage status to provide competitive advantage over other tourism attractions;
- *Operational costs* – May require investment upfront (e.g. human resources, technology) but evidence suggests long-term savings can be gained by developing sustainable marketing and management systems;
- *Positive community relations* – Essential for partnership development and local community participation. They are also valuable for building positive public relation (PR), for identifying community opportunities (participatory) and potential benefits, increasing community-based interest and value for site, facilitating community-based investments (e.g. conservation);
- *Avoidance of negative PR and achievement of positive PR* – Negative PR can inevitably lead to loss of customers and also threaten heritage integrity/status. Negative PR can be avoided by pursuing the sustainable marketing approach;
- *Regulatory compliance* – In addition to heritage status requirements, an increasing number and complexity of regulations and protection measures are being applied (laws, regulations, voluntary codes of conduct, accreditation, and award schemes). Meeting standards ahead of compliance may confer competitive advantage internationally, nationally, and locally. Integrating sustainability into management planning facilitates compliance from the onset and communicates these values to stakeholders; and
- *Preservation/conservation* – Sustainable marketing facilitates natural and cultural heritage preservation through ethical standards, long-term planning, transparent communications, best-practice reporting, and visitor education programs (Donohoe et al. 2016, pp. 6–8).

It is argued that sustainable marketing requires investment and a budget to gather insights on the market and its impacts on the natural, cultural, and economic environments so that operational veracities can inform planning of sustainable marketing and long-term cost effective strategies. Donohoe et al. (2016) emphasize on the important role of heritage tourism managers and call for the development of a strategic research agenda to assess outcomes of visitation and partnerships so that environmental, social, and economic impacts of heritage tourism can be monitored in an ongoing manner. In a more recent study, McCamley and Gilmore (2018) develop a conceptual heritage tourism marketing model and base it on four core functions: strategic orientation, resource allocation, product service development, and destination promotion (2018, p. 2). The authors call for a multi-sector integrative effort involving the government, private sector, and local community. Horizontal coordination is specifically emphasized and the role of the government is considered crucial to the success of 'policy implementation'.

An innovative optimizing framework is developed by Fauconberg, Berthon, and Berthon (2018). Their aesthetics and ontology model is aimed at forging strong bonds between visitors and international heritage attractions such as the WHS. Fauconberg et al.'s work is premised on Popper's three-world hypothesis based on luxury artifacts (as WHS are regarded as nonessential luxury goods): the physical (the intangible/physical form), the subjective (psychological experience of a person with the artifact or site and the intersubjective (the collective, socially constructed world of the artifact – such as theories, stories, myths, and social institutions); each sphere becomes whole only after connecting with the other spheres (2017, p. 3). The WHS is presented in the context of the three realms: the physical site, the topographic features, and the complete ecosystem in the context of conceptualizations, narratives, myths, stories, and social institutions. Although most preservation initiatives are directed toward the tangible form of the site, Fauconberg et al. argue that equally vital is the atmospherics (the site's ambience and spirit). In their proposed framework, they employ two dimensions: aesthetics (novice versus expert) and ontology (endurance or the unchanging versus being transient or in process) and present four modes of luxury consumption:

- *Commercial conspicuous possession* (Modern) – this ontological mode stresses on endurance. This mode has a low barrier of entry: money is used to access a world of commercialized luxury where expertize is not needed. This mode is exemplified by the Gucci bag;
- *Monumental aesthetic possession* (Classic) – this ontological mode also embraces the 'enduring' albeit with an expert eye for the aesthetic. It is about educating the eye to see that 'special quality'. As such, money, experience, and expertize are necessary investments to appreciate this mode. For example, unlike an enthusiast, the novice is unable to appreciate the superb balance and craftsmanship of a Holland & Holland Royal side-by-side;

- *Evanescent conspicuous consumption* (Postmodern) – this ontological mode embraces the 'transient' notion and reverts to a novice aesthetic. It can be described as 'conspicuous consumption' and highlights obsession with glitz, glamour, and the hyper-real. Knowledge is seen as a hindrance and barrier of entry. The mode is epitomized by such creations as the Venetian hotel in Las Vegas: it has all the wonder of Venice (canals, gondolas, and buildings), without the inconveniences of the real (garbage, smell, and mosquitoes); and
- *Ephemeral aesthetic consumption* (Wabi Sabi) – Luxury, here, is 'the ephemeral – the rare orchid that blooms for just one day ... lies in the deep taste of the moment: it is mindfulness of ephemerality. As with the 'Classic' mode, money, experience, and expertize are all necessary albeit with a shift in ontological frame. This model is embodied in the British obsession with wild gardens (Berthon, Pitt, Parent, & Berthon 2009, pp. 51–53).

According to Fauconberg et al. (2018), each mode attracts a distinct consumer market. For instance, the 'Modern' category seeks commercial experience at the site; the 'Classic' types go for monumental aesthetic appeal. They seek behind-the-stage and objectively authentic experiences. The 'Postmodern' types seek evanescent conspicuous consumption and are attracted by the pseudo and romanticized facets of the site. They show appeal for high-branded products and experiences. The 'Wabi Sabi' types prefer ephemeral aesthetic consumption and seek optimized experiences by placing faith in narratives authenticated by legitimized sources and experts.

In terms of its application value, the priority is to maintain harmony between consumer needs and conservation of the site and its peripheral areas. Second, it can be argued that partnering with ancillary heritage tourism sectors in the peripheral areas promotes sustainability and helps to offer a complete and satisfactory visitor experience. Last, the authors point out that both effective marketing and management of these sites accentuate potential for the sites to be self-sustaining in terms of economic numerations and conservation. Effectiveness of sustainable marketing initiatives relies on time investment, funding, and integration with other core operations.

In terms of featuring authenticity in the aforementioned marketing models, it is interesting to note that for the most part, its presence remains either invisible or disguised under the conservation label. The latter observation implies that, by stressing on the quality and conservation of heritage, the marketing initiatives are indirectly safeguarding the original features (objective authenticity) of a heritage attraction or a site. What has emerged in the recent decade is its elevation as a stand-alone asset and initiatives to optimize its potential using numerous branding strategies. Authenticity, on its own, has emerged as a keystone of contemporary marketing strategies that extend beyond the realm of tourism. That said, today, it is popularly recognized as an important branding tool because it is noted to be "a key purchasing criterion in heritage tourism" (Chhabra & Kim 2018, p. 55). Authenticity stimulates market growth by building emotional bonds and significantly impacts consumer behavior in a positive manner (Napoli et al. 2014). In the next section, insights are offered on the manner in which branding literature continues to elevate authenticity and leverage its potential to attract consumer markets.

Authenticity and its branding

Authenticity is receiving unprecedented attention in branding literature as brand authenticity and as a core expression of brand heritage. As consumer quest for unique and cultural experiences continues to surge (Chhabra 2010; Timothy 2011), importance of authenticity as a branding tool is no surprise (Kolar & Zabkar 2010). Particularly, authenticity perceptions influence the visitor decision-making process and can offer insights on how visitor satisfaction is influenced by authentic experiences.

Brand heritage refers to a distinct marketing strategy containing a fusion of linked brand narratives that can serve to "evoke images of continuity, safety, and familiarity amongst consumers offering reassurance and stability" (Taheri, Farrington, Curran, & O'Gorman 2018, p. 52). Taheri et al. (2018) write that a significant relationship exists between brand heritage, cultural motivation, authenticity, relational value, and commitment (loyalty). It is posited that brand heritage can strengthen consumer perceptions of objective authenticity. Brands predate on human subjectivity and focus on perpetuating audience views of value throughout the experience journey (Wuestefeld, Hennigs, Schmidt, & Wiedmann 2012) to develop relational values thereby forging an emotional bond between consumers and the employees. In the context of negotiated authenticity, relational value happens as consumers experience authenticity in a manner that it resonates with their preconceived notion of brand heritage. A significant relationship is noted between existentialist authenticity and brand heritage. That is, preconceived image of brand heritage can stimulate personal/optimally authentic experiences at the site. Therefore, "understanding these links is vital for professional marketing strategies", especially for modified sites, as an existentialist type of experience for visitor can be designed with the use of creative tools. Brand heritage needs to be effectively communicated to stimulate authenticity perceptions at all three travel phases: before visit, at the site, and post visit (Taheri et al. 2018, p. 62).

A consumer-based model of authenticity is proposed by Kolar and Zabkar (2010). It is underpinned on the notion that "authenticity as an evaluative judgment may additionally enrich the understanding of tourist experiences and behaviors and serve for marketing management purposes" (Kolar & Zabkar 2010, p. 655). They test their model and report that cultural motivations shape both objective and existential anticipations of authenticity which in turn influence the type of experiences sought by heritage tourists.

Brand authenticity holds tremendous potential to draw lucrative markets. It can be defined as the perceived genuineness of a brand expressed in terms of its continuity, ingenuity, trustworthiness, and representation (Chhabra & Kim 2018; Napoli et al. 2014). In examining brand authenticity of a heritage festival hosted by a Diaspora community, Chhabra shares that:

> the organizers leverage the festival's brand authenticity by showcasing continuity, genuineness and aligned with literature, object authenticity is orchestrated through cultural dance performances, music, garb, ... and traditional food. The diaspora is more driven by push motivations as they seek nostalgic and socializing opportunities. They are more likely to seek authenticity of the self as they seek optimized experiences and ... social/familial ties. Pull motives are not a big draw for them as they are, apparently, familiar with the commodified nature of the festival offerings. On the other hand, the non-diaspora markets are motivated by pull factors ... Furthermore, high levels of perceived authenticity, among the non-diaspora, indicate that the authenticness of festival offerings is seldom questioned. The festival's brand authenticity appeals to ... ... the diaspora and the non-Indians but in a different manner (2018, pp. 56–57).

Chhabra's authenticity branding plan is presented in Figure 5.2. She argues that the object-based elements should be showcased and their story narrated in a manner that they stimulate relaxation, instinctive encounters, and a feeling of connectedness. Theoplacity and the other negotiated demand experiences (between intrinsic and the manufactured/pseudo) can be offered at a heritage site as a packaged bundle containing attributes such as tangible traits, atmospherics, interpretation, and adaptive history to

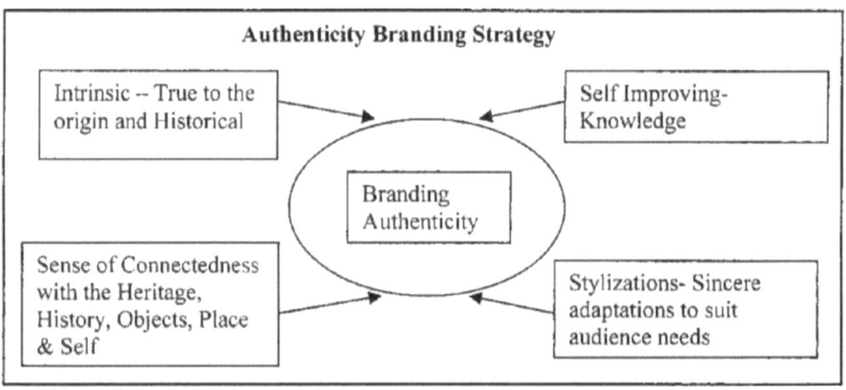

Figure 5.2 Authentic branding strategy.

Source: Chhabra (2010, p. 739)

resonate with the captivating audience so that strong connections can be forged. The design of the brand can be reliant on inherent/genuine uniqueness, fusion of actual and staged presentations that synthesize aforementioned authentic elements; the purpose is to facilitate educational/cognitive satisfaction and engulf a sense of connection with the setting's heritage, people, past history, and physical elements. It can appeal to distinct demand for authentic experiences:

> With an aim to stay above commercial considerations that are threatening the heritage industry because of financial constraints. The proposed measures can exemplify real commitment to conservation of the true nature for the heritage site and its narratives. The brand mix consists of key considerations such as use of the site as a referent, commitment to conservation of the tangible and intangible elements at the site, use of historians and contemporary culture as referents to create messages that can be inspire contemporary audience and reinforce links to the past (Beverland 2005). In this manner, the heritage sites can keep themselves in check while practicing commercial marketing strategies such as branding. As suggested by Beverland (2005), this type of authentic branding strategy can downplay actual scientific and marketing prowess and conform to the expected rules of conservation, showcase intrinsic, historical continuity, and demonstrate conservative attitude to change. (Chhabra 2010, pp. 739–740)

Chhabra's study illustrates the significance of authenticity as an important performance indicator of heritage tourism and management of cultural heritage. Authenticity, as a means for appraisal, can gauge favorable tourist experience and satisfaction and can be an important indicator for measuring marketing effectiveness. Moreover, branding of some forms of authenticity will need an ongoing interactive dialogue as markets change and search for an authentic self continues to develop. Evidently, satisfaction can be influenced by different types of authenticities. Both existentialist and objective types of authenticities are popularly sought and they augment satisfaction levels. But it is important to point out authenticity needs to be strategized and is but one crucial element of a bigger tourism system. I further this argument in the forthcoming paragraphs by integrating authenticity into a systems framework.

Authenticity and a tourism systems perspective

Limited studies have incorporated a harmonious mix of marketing, authenticity, culture/heritage, commodification, stakeholder collaborations with special focus on local community involvement, and conservation. A smart view of sustainable of marketing advocates integration of multiple components such as: ICT, brand personality, ethical consumption, ethical production, corporate governance, and brand personality (Chhabra 2015).

It is important for marketing to employ a tourism systems framework adaptable to heritage tourism settings (McKercher 1999; Mill & Morrison 2006). A system's perspective offers a long-term view, supports economic, and strengthens supply chain linkages thereby generating value both for the suppliers as well as the consumer markets. Tourism system is a dynamic mix of multifaceted elements that are constantly in a state of chaos and disequilibrium. Several models are presented in documented literature. For instance, Leiper's (1990) tourism system model takes into consideration the complex character of tourism. The author employs a theoretical approach to make sense of the complex nature of tourist attractions. According to him, "a tourist attraction has nuclear parts which extend beyond the threshold of its physical aspects. Tourists make the attraction a tourist attraction, the areas in the immediate vicinity of nucleus represent the space ...... to reach the core, and the different markers act as pieces of information. Markers can be both off-sight and on-sight and can be of various kinds such as detached experiences, and help attach meaning to the touristic experience" (1990, p. 378). Worthy of note is McKercher's tourism system model:

> It argues that things do not work as planned as there may be several intervening factors at play to upset the equilibrium state. Here, he calls to attention the significance of embracing the chaos and complexity theory to factor unpredictability into the tourism system equation. Arguing that even stable tourism communities and systems experience turbulence and keeping this unpredictability in view, a chaos-centric tourism functioning model is proposed. Chaos theory, in this case, implies disturbance, "loss of control," or a "situation where a system is dislodged from its steady state after being conditioned by a catalyst that is as random and unpredictable as the outcome" (Russell 2006, p. 167). This means that tourism systems are dynamic and constantly changing. That said, chaotic systems also have the ability to self-organize themselves through adaptive strategies, if initiatives are taken. Some pronounced effects in the chaos context include the 'butterfly' and 'lock-in' effects. The model draws heavily on the principles of chaos and complexity. It is an open model with unrestricted mobility of traveler flow to the outputs or supply side of the system. It has nine inter-connected key elements: 1) the traveler, 2) the communication vectors, 3) the considerations, 4) the destination or Internal tourism community, 5) the external tourism agencies, 6) other tourism-related externalities, 7) non-tourism-related externalities, 8) outputs from the system, and 9) and rogues or Chaos makers. While the number of actors influences the system changes at each level, the relationships between elements remain similar, and thus the model continues to work. Tourism functions in a non-linear, non-deterministic and dynamic manner and turbulence and periods of intense upheaval are both an intrinsic element of the system and essential element to promote rapid change in tourism communities. (1999, pp. 428–433)

McKercher's model can be extended to develop an adaptive 'Complex Tourism Systems' marketing paradigm to facilitate innovative insights/solutions by adding ethical consumption, value added centricism, and interactivity (given the unprecedented popularity of online marketing initiatives). Multi-tiered supply chain linkages and synergies between different tourism offerings are crucial. In the context of smart marketing of heritage tourism, Chhabra (2015) argues that adaptive and strategic use of technology is required and it needs to be leveraged in a manner that it complements/strengthens the core essence of brand authenticity by optimizes negotiated authentic experiences. The adaptive complex tourism systems approach leverages technology to add value and enrich authentic heritage experiences in a sustainable and equitable manner.

Closing comments

Although for many decades, heritage practitioners, across the globe, have been critical of marketing and regard it as a hindrance to conservation efforts, recent literature argues that "marketing holds many potential solutions to some of the challenges around sustainability" associated with heritage tourism (Gordon, Carrigan, & Hastings 2011). Marketing efforts should be aligned to the overarching purpose of conserving objective authenticity by striking a delicate balance between commodification and the original value of a heritage resource (tangible and/or intangible). Negotiated authenticity versions can be strategized and branded using holistic sustainable marketing indicators, grounded in a tourism systems framework.

Undeniably authenticity has emerged as a powerful branding tool and "an important performance indicator. Authenticity, as an evaluative judgment, is a crucial measure to gauge marketing success. Strategic branding of authenticity calls for an interactive dialectical process as markets change and quest for self continues to exponentially grow. The fluid nature of the negotiated authenticity need will continue to test the marketing prowess of heritage institutions as they strive to strike a delicate balance with the intrinsic values of heritage products" (Chhabra 2010, pp. 739–740).

In some conservation scenarios selective marketing, demarketing, and/or social marketing techniques might be more relevant. For instance, demarketing can be employed by integrating marketing strategies with visitor management techniques (Beeton & Pinge 2003). In this regard, a specific threshold needs to be identified where visitors are not permitted or their activities are restricted. Undoubtedly, much burden to sustainably manage visitors and their impacts stems from the need to make heritage plausible, worthy, and meaningful to visitors. Notable strategies to integrate visitor management and marketing are offered by Fullerton et al.:

> Influencing visitors at the decision making stage. Marketing at this stage can be directed towards a target audience to influence behavior.

> Demarketing can be used to discourage certain types of visitors. By including demarketing in the marketing mix, a destination may attract environmentally aware visitors and select specific markets, thereby enforcing above two types of demarketing. The marketing agenda needs to include … … … … tools such as … educating potential visitors, encouraging preferred markets and discouraging unappealing ones, publicizing alternative routes, limiting potential activities either seasonally or entirely, and limiting access to fragile areas and divert visitor traffic to unfragile areas;
>
> Incorporating/integrating environmental and societal perspectives into the sustainable marketing paradigms focused on producing viable economic benefits for the host community;
>
> Greater professionalism and emphasis on marketing research to design strategies relevant to the contemporary environment and preferred target markets;
>
> Effective interpretation techniques and programs to guide and manage visitor behavior. (2018, p. 112)

Key here is a smart heritage tourism systems approach underpinned on negotiated authenticity that aims to authenticate objective authenticity and optimizes consumer experiences while generating viable socioeconomic benefits for host communities. This approach can feature a mix of sustainable, demarketing and social marketing strategies to mobilize negotiated authenticity and make its authentication functional and culturally resilient to predicable and unpredictable disturbances in the heritage environment. These recommendations should also hold value during post-covid times, although long-distance heritage travel might not become popular for some years. People are more likely to travel closer to their homes and this shift will give impetus to more localized heritage tourism and opportunities to reform commodified heritage in a manner that it can hold a mirror to its objectively authentic essence.

References

AMA. (2007). *Definition of marketing*. Retrieved: httep://www.marketingpower.com/AboutAMA/Pages/DefinitionofMarketing,aspx

Beeton, S., & Pinge, I. (2003). Casting the holiday dice: Demarketing gambling to encourage local tourism. *Current Issues in Tourism*, *6*(4), 309–322.

Berthon, P., Pitt, L., Parent, M., & Berthon, J. (2009). Aesthetics and ephemerality: Observing and preserving the luxury brand. *California Management Review*, *52*(1), 45–66.

Beverland, M. (2005). Crafting brand authenticity: The case of luxury wine. *Journal of Management Studies*, *42*, 1003–1030.

Butler, G., & Ivanovic, M. (2016). Cultural heritage tourism development in post-apartheid South Africa: Critical issues and challenges. *Cultural Tourism in Southern Africa*, 58–75.

Chhabra, D. (2009). Proposing a sustainable marketing framework for heritage tourism. *Journal of Sustainable Tourism, 17*(3), 303–320.

Chhabra, D. (2010). Branding authenticity. *Tourism Analysis, 15*(6), 735–740.

Chhabra, D. (2015). A cultural hospitality framework for heritage accommodations. *Journal of Heritage Tourism, 10*(2), 184–190.

Chhabra, D. & Kim, E. (2018). Brand authenticity of heritage festivals. *Annals of Tourism Research, 68*, 55–57.

Chhabra, D., & Zhao, S. (2015). Present-centered dialogue with heritage representations. *Annals of Tourism Research, 55*, 94–109.

Coccossis, H. (2008). Cultural heritage, local resources, and sustainable tourism. *International Journal of Services, Technology and Management, 10*(1), 54–60.

Donohoe, H. M. (2012). Sustainable heritage tourism marketing and Canada's Rideau Canal world heritage site. *Journal of Sustainable Tourism, 20*(1), 121–142.

Donohoe, H. M., & Needham, R. D. (2009). Moving best practice forward: Delphi characteristics, advantages, potential problems, and solutions. *International Journal of Tourism Research, 11*(5), 415–437.

Donohoe, H. M., & Needham, R. D. (2008). Internet-based ecotourism marketing: Evaluating Canadian sensitivity to ecotourism tenets. Journal of Ecotourism, 7(1), 15–43.

Donohoe, M., Salo, B., & Gilmore, T. C. (2016). The Promise And Potential Of Sustainable Heritage Tourism Marketing.

Fauconberg, A., Berthon, P., & Berthon, J. (2018). Rethinking the marketing of world heritage sites: Giving the past a sustainable future. *Journal of Public Affairs*, 18e. https://doi.org/10.1002/pa.1655.

Font, X., & McCabe, S. (2017). Sustainability and marketing in tourism: Its contexts, paradoxes, approaches, challenges and potential. *Journal of Sustainable Tourism, 25*(7), 869–883.

Gilmore, A., Carson, D., & Ascencao, M. (2007). Sustainable tourism marketing at a world heritage site. *Journal of Strategic Marketing, 15*, 253–264.

Gordon, R., Carrigan, M., & Hastings, G. (2011). A framework for sustainable marketing. *Marketing Theory, 11*(2), 143–163.

Hargrove, C. (2003). Authenticity in cultural heritage tourism. *Forum Journal: The Journal of the National Trust for Historic Preservation, 18*(1), 45–52.

Hede, A. M., & Thyne, M. (2007). *Authenticity and branding for literary heritage attractions*. Australia and New Zealand Marketing Academy.

Hunter, C. (1997). Sustainable tourism as an adaptive paradigm. *Annals of tourism research, 24*(4), 850–867.

Jamieson, W. (2000). *The challenges of sustainable community heritage tourism, heritage management and tourism conference*. Bhaktapur, Nepal: UNESCO Workshop on Culture.

Kolar, T., & Zabkar, V. (2010). A consumer-based model of authenticity: An oxymoron or the foundation of cultural heritage marketing? *Tourism management, 31*(5), 652–664.

Leiper, N. (1990). Tourist attraction systems. *Annals of tourism research, 17*(3), 367–384.

McCamley, C., & Gilmore, A. (2018). Strategic marketing planning for heritage tourism: A conceptual model and empirical findings from two emerging heritage regions. *Journal of Strategic Marketing, 26*(2), 156–173.

McKercher, B. (1999). A chaos approach to tourism. *Tourism Management, 20*(4), 425–434.

McLean, F. (2002). *Marketing the museum*. London: Routledge.

Mill, R. C., & Morrison, A. M. (2006). *The tourism system*. Kendall Hunt.

Misiura, S. (2006). *Heritage marketing*. Boston, MA: Elsevier.

Moscardo, G. (2001). Cultural and heritage tourism: The great debates. In B. Faulkner, G. Moscardo, & E. Laws (Eds.), *Laws tourism in the 21st century* (pp. 3–17). London: Continuum.

Moscardo, G., & Woods, B. (1998). Interpretation and the experience of visitors on Skyrail. Embracing and managing change in tourism: international case studies, 311.

Mowforth, M., & Munt, I. (2016). *Tourism and sustainability: Development, globalisation and new tourism in the third world*. London: Routledge.

Napoli, J., Dickinson, S., Beverland, M., & Farrelly, F. (2014). Measuring consumer-based brand authenticity. *Journal of Business Research*, *67*, 1090–1098.

Parsons, E., & Maclaran, P. (2009). *Contemporary issues in marketing and consumer behaviour*. Routledge.

Phelan, A., Ruhanen, L., & Mair, J. (2020). Ecosystem services approach for community-based ecotourism: Towards an equitable and sustainable blue economy. *Journal of Sustainable Tourism*, *28*(10), 1665–1685.

Rowan, Y., & Baram, U. (Eds). (2004). *Marketing heritage. Archaeology and the consumption of past*. Walnut Greek, CA: Altamira Press.

Russell, R. (2006). 10. Chaos Theory and its application to the TALC model. In *The tourism area life cycle* (Vol. 2, pp. 164–180). Channel View Publications.

Taheri, B., Farrington, T., Curran, R., & O'Gorman, K. (2018). Sustainability and the authentic experience. Harnessing brand heritage–a study from Japan. *Journal of Sustainable Tourism*, *26*(1), 49–67.

Timothy, D., & Boyd, S. (2003). *Heritage tourism*. New York, NY: Prentice Hall.

Timothy, D. J. (2011). *Cultural heritage and tourism: An introduction* (Vol. 4). Channel View Publications.

Wuestefeld, T., Hennigs, N., Schmidt, S., & Wiedmann, K.-P. (2012). The impact of brand heritage on customer perceived value. *Der markt*, *51*(2–3), 51–61.

6 Negotiated authentication of heritage accommodations

This chapter offers insights on the manner in which the heritage hotels orchestrate negotiated authenticity. It scrutinizes the political and economic discourses that shape their authentication process. A discursive view is taken to identify factors that initiate and contribute to their vulnerability so that a proactive authentication strategy can be proposed to preserve the core essence of the heritage properties.

Heritage hotels and resorts are an important subject of deliberations in the authenticity discourse because they are valorized for offering culturally authentic experiences. This can be evidenced from a review of promotional themes of many heritage properties across the globe. As an instance, Heritage Hotels & Resorts, Inc., in the United States promises its guests: an authentic Southwestern experience in New Mexico's best tourist destinations. Through our architecture, interior design, original artwork, landscaping, entertainment, and cuisine, Heritage Hotels & Resorts, Inc., provides guests with an authentic, cultural experience in Santa Fe, Albuquerque, Taos, and Las Cruces.

The museum hotel, Auberge Saint-Antoine in Quebec, promotes itself along similar lines: *Anchor your Quebec City experience with a stay at Auberge Saint-Antoine, where our warm, convivial atmosphere greets each guest upon arrival. Throughout your stay, be treated to authentic hospitality, boutique accommodations, farm-to-fork dining, out of the ordinary event space and more, while surrounded by the old-world, yet modern charm of our historical Quebec hotel. Welcome to Auberge Saint-Antoine, where history and hospitality blend to create a memorable hotel experience.*

Undeniably, one way to appropriate cultural and historical markers, for bygone eras, in heritage tourism is adaptive reuse of old buildings. Refurbished hotels or remodeled historic buildings have become a vital part of the heritage tourism portfolio in different countries; they are regarded as an important marketable commodity (Gholitabar & Costa 2018; Henderson 2013; Timothy & Teye 2009; Xie & Shi 2019) because they enhance the cultural attractiveness of a given destination. Such built structures also hold economic value (by generating direct and indirect revenues), enriching rural/city life and urban landscapes, and add to place novelty (Ebbe 2009; Langston, Wong, Hui, & Shen 2008; Pongsempol & Upala 2017). For instance, historic hotels boost the contemporary market demand for a nostalgic past, cultural gastronomies, and slow authentic food experiences. In compliment, such unique properties have become a part of the 'International Hotels Environment Initiative' devoted to sustainability-related issues associated with eco-efficiency, cultural management, ecolabels, and ethical practice guidelines.

In fact, support for heritage hotels is part of a bigger initiative in many countries, where the overarching aim is to promote conservation (such as preservation and restoration) of old historic structures. Xie and Shi write that "historic hotels, heritage lodgings, and other such tourist accommodations, involving gentrification process, have become an important part of preservation and promotion in any given destination" (2019, p. 67). Historic buildings in local regions, reframed and repackaged in the form of heritage hotels, can optimize local benefits by:

1 their long-term orientation and stability;
2 by involving the local community in the provision of ancillary tourism services; and
3 forging links and strengthening dependencies between local heritage-related activities (Murzyn-Kupisz 2013, p. 157).

Because of their economic value, heritage hotels have captured the interest of the private and public sectors. Additionally, as mentioned above, they have earned the support of local communities by supporting and showcasing their conspicuous cultural/heritage attributes. Therefore, heritage hotels as a tourism commodity have many uses:

- They can be used as a tool to promote civic pride, local identity, and cultural capital.
- They promote heritage sustainability, especially if their heritage attributes are endorsed by the local community and the inculcated value drawn from their traditional knowledge (Greffe 2004).
- They can help bridge the gap between the locals and the tourists by projecting an "attractive image for investors, tourists, and local residents" (Chang 1997, p. 47).
- They can help preserve remains of "original structures or historical artifacts. Also the preservation of traditional forms of lodging also contributes to retain the cultural geography of a region" (Aslam & Joliffe 2015, pp. 121, 126).

Based on a review of documented literature, this chapter offers insights on the manner heritage hotels orchestrate negotiated authenticity. Furthermore, it scrutinizes the political and economic discourses that shape the authentication processes. A discursive view is taken to identify factors that make such heritage structures vulnerable. The purpose is to propose a proactive authentication strategy to preserve the core essence of these heritage properties. Three core questions guide the purpose of this chapter:

- How is authenticity negotiated to preserve the history and heritage of hotels?
- How is the preferred version of authenticity authenticated?
- What type of factors shape, support, or compromise the authentication process?

Authenticity standpoint

Most studies offer a supplier's perspective and illustrate advocacy for the negotiated version of authenticity. For instance, de Klumbis and Munsters (2005) refer to historic properties as a lifestyle commodity with augmented experiential value, high quality of personalized service, and cultural gastronomy. Heritage hotels mix yesteryear style with modern day amenities to elevate the market demand (Henderson 2011; Lee & Chhabra 2015). Aslam and Joliffe (2015) illustrate how built heritage, with the colonial tea industry, has been repositioned as heritage accommodations for tourists in Ceylon. Their study focuses on the tea heritage and shares the significance of historic tea lodgings in their potential to promote heritage tourism. The authors reiterate demand for novel and negotiated authentic experiences to meet the contemporary tourist demand for allocentric experiences. The authors describe that: "historic homes are set on a large block of land, and have either a colonial or esthetic architecture. However, restoring or establishing historic accommodations in small estate lands and refurbishing as per the operator's or proprietor's wish without regulatory guidelines reveals lack of strong policy and implementation of standards. This, coupled with increasing demand for historic lodging mystifies the tea culture and heritage, leading to staged authenticity" (Aslam & Joliffe 2015, p. 120).

It is also important to note that objectively authentic versions of heritage and history might not be pleasing to the eye; that is, reality can be uncomfortable to experience, both emotionally and from a tangible standpoint. From a tangible viewpoint, Aslam and Joliffe report numerous operational issues with heritage accommodations such as: ensuring good service quality and guest service and "limited capacity of units" due to the true "nature of the historic structures and their maintenance"; training the local community also poses a challenge coupled with limited resources to pursue "branded and segmented tourist product marketing" and poor transportation infrastructure (2015, p. 121). Such built heritage resources require adaptative strategies to make them user friendly. Therefore, the key test lies in the commodification (compromise) of objective authenticity in a manner that the value and core of original historic structures remain intact.

Chhabra (2010) examines a heritage resort on an Indian reservation (the Sheraton Horse Pass Resort) in the United States. She reports that its culture and heritage representations are drawn from a mix of objective (confirmed through murals, tribal language, and handicrafts) and negotiated perspectives of authenticity. Negotiated authenticity is pursued in the use of materials such as wood. Today, the woodwork tradition has replaced the traditional farming and basket weaving practices because of the damming of the river that runs through the reservation. The waters of the Gila River have been rerouted to support the non-tribal farmers. Although the Indians have adapted to this change which had shaped their cultural lifestyle, they continue to ensure that their modified practices resonate with the core cultural values of the tribe. For instance, painted ceiling murals visually authenticate narratives which harmoniously portray negotiated authenticity of the tribal culture (see Figures 6.1 and 6.2). It is

Figure 6.1 Sheraton Grand at Wild Horse Pass Resort.

Figure 6.2 Murals – Sheraton Grand at Wild Horse Pass Resort.

further noted that: "both preservation and regeneration techniques are employed to support objective authenticity. Some artifacts are displayed and preserved in their original condition. For instance, a calendar stick is on display, in a glass case, with an information tag stating: "prior to written language, all Pima and Maricopa historical events and legends were passed down orally. The only written history was recorded on a calendar sticks such as these. Each notch represents a time frame and each symbol represents a significant occurrence that took place during that time" (Chhabra 2010, pp. 133–134).

Adaptive reuse of buildings as heritage hotels is also growing in other parts of the world, such as Southeast Asia. Examples include Raffles Hotel and Fullerton Hotel in Singapore, Umaid Bhawan in Jodhpur (Rajasthan) in India (Table 6.1), a hotel in Nepal and Myanmar. Raffles Hotel was built during the British rule while Fullerton Hotel used to be an administrative office during the colonial era. Boutique iconic hotels have also become an important subset of heritage hotels. Key features of a boutique iconic hotel are: its personalized service; it is not connected to a large hotel chain; it has less than 100 rooms and; it is historically and culturally unique (Henderson 2011). From an authenticity perspective, this category of niche hotels showcases the constructivist version of authenticity although the aim is to embrace the negotiated stance by integrating elements unique to objective authenticity. This uniqueness is visible in the integration of "striking modern architecture or buildings and settings of historical significance" (Chang & Teo 2008). In fact, this category constitutes a new generation of hotels with "low key historical authenticity"

Table 6.1 Cultural hospitality of Umaid Bhawan (heritage resort in India)

• Umaid Bhawan Palace is a popular attraction of Jodhpur and one of the largest royal residences in the world. Taj Hotels is managing a part of the palace since 1972. The Palace is divided into three functional parts: the luxurious Taj Palace Hotel, the residence of the present king of Jodhpur, and a museum exhibiting the history of the city tracing the royal family and its lineage to the twentieth century. It also shares history of the contribution of the royal family in building and developing the city of Jodhpur. The palace is rooted in local community history as it was initially built for the purpose of offering to residents when a famine devastated the region (Tollotson, 2011).
• The palace houses one of the most elegant resorts (with 347 rooms) in India and stands as a witness to living heritage. One portion of the palace remains occupied by the royal family. Built in yellow sandstone, it is one of the most defining markers of 'Indianness' and the living heritage of Jodhpur. A British architect was commissioned to use Indian craftsmanship by employing Indian artisans. Almost three thousand workers were employed and it took approximately 14 years to complete the monument. It depicts an amicable amalgamate of European classical art decor and Indian art and sculptures and ornaments designed according to Rajasthani customs (Singh 2006; Tollotson, 2011).
• The palace compliments the surrounding landscape and the local traditions of the town. The ancestors of the current generation of Jodhpuris dedicated many years of their life in building the palace. As per local customs of Suryavanshi Rajput dynasties, the main entrance faces the east. Decoration bands inside the palace staircases are taken from Hindu temples. The whole ambience conveys dignity and class, conveyed to guests through delicately devised cultural hospitality codes.
• Each guest is greeted like royalty with a marigold garland, the beating of drums, addressed as 'hukum' (your highness), and is escorted by a soldier in a traditional garb, holding a sword which was used to protect kings in early times. Several oral expressions of history are evident in the use of the local language and hospitality. The royal family continues to be integrated with the local community by gracing its presence in the local festivals and other cultural events. It plays an active role in advocating for community issues, particularly related to infrastructure and water supply. Most of the staff in the hotel are locals.

Source: Chhabra (2015).

(Chang 1997; Keeps 2006). An iconic boutique hotel is based on adaptive principles, as illustrated by Henderson:

- According to an American independent hotelier overseeing the conversion of a 1929 Hollywood Art Deco landmark, the 'trick' lies in reinterpreting a building's history into 'something that works now' while connecting the guests to the past (Keeps 2006). One approach is to incorporate contemporary or futuristic furnishings and fittings into the historic structures using a fusion of past, present and future (de Klumbis & Munsters 2005);
- The success of boutique hotels would thus seem to depend upon defining attributes of smallness, service and independence. Individuality is an essential dimension based on site and property details which are able to exercise sustainable heritage appeal (2011, p. 219).

Figure 6.3 Cultural landscape of Umaid Bhawan.

Chhabra also scrutinizes the notion of authenticity using a cultural hospitality framework for a heritage resort, Umaid Bhawan (a high-end resort in Rajasthan, India). Table 6.1 highlights the uniqueness of the heritage property. Figures 6.3 and 6.4 share the manner in which 'heritageness' is authenticated at the resort. Negotiated authenticity is portrayed through traditional service offered to the guests. Cultural hospitality is a term that mirrors the attributes of negotiated authenticity because it is derived from the cultural codes of the host community and the uniqueness of the historic setting. It advocates cultural authenticity in objective or negotiated forms. It is contented that, at its very core, a sustainable cultural hospitality framework should:

1 showcase past history of the built heritage and its embeddedness within the local community;
2 generate a sense of pride within the local residents and garner their support;
3 support cultural practices/norms of the community;
4 provide economic benefits to the community through employment and other integrated offerings; and
5 offer a long-term strategy to strengthen the sustainability of cultural/heritage in the local environment (Chhabra 2015, p. 186).

The aforementioned account and Table 6.1 illustrated delicate manifestation of essentialist authenticity in the heritage experience package. The local

Figure 6.4 The heritage room of Umaid Bhawan.

culture and the royal heritage are deeply integrated into the resort environment and the guest experiences. The traditional hospitality, evidenced in the welcoming of the resort guests by the beating of drums and flowery garlands and uniform of the staff, illustrate some of the culturally inspired initiatives of the resort management. As another instance, the guest experience is authenticated with historical narratives and the resort features such as the traditional layout of its landscape and architecture. Objective authenticity is negotiated in a delicate and appealing manner.

Elaborating on cultural hospitality codes, based on the interviews with the resort staff, Chhabra further reports:

> Jodhpuri cultural norms and traditions are embraced in the resort's hospitality agenda. The staff is educated about different local folk dances and music (such as flute playing) so that they can inform guests

> and culturally immerse them insofar as possible. Sustainability principles are promoted by inviting guests to engage with local vegetation and wildlife. The peacock is the cultural symbol of Rajasthan, and pictures and engravings of the bird are evident in architectural designs. Guests are given guided tours of the palace, its gardens, and are informed of the manner they should approach the peacocks, which wander freely in its gardens. In a nutshell, this heritage property is distinct and culturally embedded in local traditions. Existing documents and staff conversations point to its past history and its chronological distance – 80 years of existence. Staff training includes imparting knowledge of the palace, its architectural history, and information on different exhibits displayed in the lobby and its corridors. This is entwined with the history of the royal family's traditions/lineage. For instance, most portraits belong to kings, not queens, because of the purdah system. Pictures of women were not publicly displayed in the bygone era. With regard to community embeddedness, continued support by the palace of local events is evident in the manner the local festivals are promoted to guests such as the kite-flying festival and the Pushkar festival. These initiatives offer evidence of integrated heritage offerings and economic linkages with the city (2015, pp. 188–189).

The Havelis in Rajasthan (India) are another example of negotiated cultural expressions. The traditional Haveli includes the following core elements/components: Courtyard (serves many purposes as a center for various rituals & household activities, and place of worship of the tulsi plant); Parsal (Tibara), is a shaded place with low intensity of light. It offers an attractive setting for both 'active' and 'passive' activities and works as a 'transitory' place for relaxation; Khadki (Choubara), is a fully shaded place that works as a filter to street heat, light, and noise; Otala is the outer place of dwelling and works as a meeting or socialization place for the family or neighbors particularly, during the winter afternoons and summer evenings. An adaptive use of these extended traditional houses has been the only way to save them from decaying, and other contemporary issues. For instance, today these Havelis are being threatened by congestion and pollution. Key issues are noted to be:

> unplanned transformation, misuse of heritage properties, traffic congestion, haphazard parking, encroachments by informal sectors, solid waste management etc. These problems may affect the unique characteristics, architectural value and heritage of the Havelis in the walled city. It is important to retain the integrity of building material and construction technology with minimum and reversible interventions. They were outstanding examples of sustainability in the hot and dry climate of Rajasthan. The havelis of Rajasthan used courtyards and other related elements as the perfect architectural response to Rajasthan's

> diverse culture and climate. But at present due to population pressure, increasing commercial activities and changing lifestyle of people, lot of transformation is taking place. Also, due to lack of maintenance, buildings are in dilapidated state and heritage is being lost. (Upadhyaya 2017, pp. 1482, 1491)

By transforming these traditional homes into heritage accommodations, they are made commercially viable (thereby optimizing economic returns). Sustainable adaptation for economic purpose is approved (by the local government) as long as the architecture of the property is not compromised and all improvements, renovations, and amendments to the existing structures harmonize with the traditional architectural style construction techniques. Such heritage hotels are popularly sought, especially, by international tourists and their authentic settings are strategically negotiated by the management, as can be evidenced from the promotional content of Alisisar Haveli hotel in Jaipur, Rajasthan (India):

> Live the Royal Rajasthani moments and regal charm with the majestic and baronial Alsisar Haveli – the heritage hotel in Rajasthan. Over the past five centuries, Maharajas of Rajasthan have enjoyed the timeless grandeur of this magnificent and stately setting, attracted by the relaxed aristocratic – feel. Handsomely located in the center of Pink City – Jaipur, Alsisar Haveli – the heritage hotel is an appealing treat of regal architecture, plush hospitality and princely ambiances. Remaining ardent to the pristine historic architecture and interior design, Alsisar Haveli – the heritage hotel has been soulfully renovated to preserve its supreme level of comfort and flair. (Alsisar Haveli 2019)

Moving forward, Henderson's work offers a dialectical view of negotiated authenticity. She notes that many of the boutique hotels "are the result of the adaptive reuse of old buildings combined with modern design to cultivate an ambience of hip heritage. The future prospects of boutique hotels with a heritage component thus partly depends on the protection of built heritage, suggesting an alignment between commercial and conservation interests" (2011, p. 217). The manner in which boutique hotels pursue negotiated authenticity in Singapore is illustrated below:

- Eight of Singapore's boutique hotels are found in the officially designated Historic District of Chinatown and adjacent streets;
- Chinatown remains a singular district in marked comparison to the internationalized and high-rise landscape of much of built Singapore. It is this factor, alongside urban planning agendas, which helps to explain the concentration of boutique hotels there;

- The special qualities of Chinatown are mined in hotel advertising with phrases such as 'rich history', 'old world charms', and 'cultural traditions of the East';
- Seven of the hotels occupy converted shophouses, which are variously described as a heritage building, 'old world architecture', and an '... old Chinatown';
- Another property, originally owned by a coolie who ... ... became a wealthy nutmeg trader, 'reflects the sentiments of the old spice merchants, depicting the typical trades of the early Chinese migrants';
- Reminders of the heritage of the hotels are largely confined to the building facades and the neighborhood, although a series of archival photographs hangs on the walls of one hotel.
- The hotelier behind the New Majestic and Hotel 1929 chose Little India (a hub of Indian culture and trade), another conserved Historic District full of shophouses, for his latest venture. Heritage is a peripheral fact of the eighth hotel, on the outskirts of Chinatown, which is a twenty-first century construction completed in 2009. It is marketed as an 'urban sculpture of unique architecture', embellished with a 'private collection of original artworks specially commissioned to reflect the hotel's flora and fauna theme' (Henderson 2011, p. 221).

Most scrutinized studies offer a supplier's viewpoint on authenticity. Literature analysis shares limited insights on type of authenticity sought by guests of heritage hotels. Of the few that can be identified, See and Goh (2019) look at tourists' intention to visit heritage hotels at George Town World Heritage Site in Malaysia. The authors use perceived authenticity as one of the predictors of visit intention along with perceived price, experience quality, prior knowledge, and social influence and report that perceived authenticity exerts significant influence on repeat visitation.

There are a few studies that discuss hospitality strategies in heritage hotels, in terms of innovative service quality models. For instance, Choo et al. scrutinize existing studies to identify several service quality models for the hospitality industry such as SERVQUAL (with five domains: tangible attributes, reliability, responsiveness, assurance, and empathy), HOLSERV (comprising of three factors: employees, tangibles, and reliability), and LODGQUAL (with five lodging quality domains: tangibility, responsiveness, reliability, communication, and confidence) (2018, p. 381). The authors argue that none of the aforementioned models scrutinizes the emotional perspective of consumers. To fill the lacuna, they have developed a holistic customer experience quality (CEXQ) model for the heritage hotels of Malaysia with four domains but offer no recognition to heritage authenticity. The authors report that customers seek a combination of 'truth of the moment' (existentialist authenticity) type of experiences in addition to valuing the experience the heritage setting offers. However, they miss a crucial element that makes it stand out: its heritage.

A complete customer quality experience model needs to add an authenticity dimension to its script. An adapted version of CEXQ is offered:

- Product experience – comprises of four metrics to stress on the need to understand what the market desires ... so that unique tailormade innovative products/experiences can be offered – freedom of choice, cross-product comparison, comparison necessity, and account management. Personalized touch-point (such as hotline or assigned contact desk) to facilitate interactive dialogue is important;
- Outcome focus – aims to facilitate purposeful and genuine experiences ... and transaction costs and is defined by inertia, result focus, past experience, and common grounding. Hotels need to ensure that the hotel processes are smooth and consistency exists by empowering employees and accurate understanding of guest needs through frequent interactions with the frontline staff;
- Moment-of-truth – relates to factors such as flexibility, ... pro-activity, risk perception, interpersonal skill, and service recovery. These highlight the significance of training employees to be proactive and successfully manage guest interactions and build long-term relationships so that service failure can be overlooked;
- Peace of mind – relates to the emotional aspects of service to positively engage customers at all touch points (including pre-purchase, during purchase, and post-purchase interactions) so that a favorable and relaxed environment is developed; and
- Authentic feel – offers a relaxing atmosphere by making the setting more objectively authentic so that differentiated and self-touch moments can be experienced. This dimension focuses on tangible and intangible heritage attributes of the heritage hotel that make it unique and distinct from the competitors. Authentic touch points can satisfy the desire for an authentic experience in a strategically commodified setting (Choo, Tan, & Yeo 2018).

By including the authentic feel dimension, the heritage attributes and authentic touch points can be emphasized to inspire customers to experience heritage authenticity in a built structure, that embodies both character and an authentic appeal. Such feelings can infuse a sense of inner self and relaxation. In other words, a harmonious relationship between service quality and cultural hospitality in adaptive heritage accommodations can promote theoplacity and existentialist authenticity.

In some parts of the globe, it is noted that hotels focusing solely on objective authenticity are more likely to attract traditional domestic markets; they may not hold appeal for international tourists. For instance, Saad, Ali, and Abdel-Ati share challenges associated with Sharia-compliant hotels (SCH) in Egypt. SCH are a form of Islamic hotels that offer services based on the Shariah code of conduct. The Muslims are required to conform their lives to the Shariah doctrine. SCH do not serve alcohol and offer Halal food, rooms for prayer, holy book of Muslims (Quran), conservative staff uniform, hire mostly Muslim staff, have a dress code for guests, ensure beds and toilets do not face the direction of Mecca (holy place of Muslims), offer prayer mats in each room, and separate amenities (such as spa, gym, and swimming pool) for men and women (Henderson 2010; Ozdemir & Met 2012). SCH popularity has accelerated especially because of the surging domestic Muslim traveler market with high-spending propensity (Rosenberg & Choufany 2009; Saad, Ali, & Abdel-Ati 2014).

Several challenges, such as loss of revenue and rating, are reported as hotels try to attract non-Muslim markets. For instance, hotels prohibiting alcohol on their premises were penalized by the Ministry of Tourism in Egypt (Sahafa 2008). Challenges include loss of revenue. Some hotels have expanded their spaces for conferences and raised the room rates to make up for the lost revenue. Capacity management is another issue as separate facilities have to be offered to males and female travelers. Cultural hospitality codes of conduct, premised on Shariah principles, have to be cautiously developed in a manner that they resonate with different stakeholders including religious organizations. The cultural hospitality of SCH is unique and might differ from other heritage hotels; therefore, it needs to be marketed in a sensitive, educational, and informative manner. It is argued that there is a growing segment of visitors who seek the genuinely authentic pulse of cultural hospitality and therefore the core of the Shariah doctrine should be kept intact while commodifying other parts of the hotel. Cultural codes need to be shaped by the host community.

In her study of heritage hotels from several countries across the globe, Henderson (2013) notes that heritage and its authentic showcasing can be viewed as a commercial resource for tourism. She shares conflicts between adaptive reuse and conservation, thereby lending credence to the negotiated version of authenticity. As corporate interest in heritage hotels continues to soar, an adaptive cultural hospitality blueprint needs to be developed to ensure that the original heritage is not distorted to the extent that it completely loses its historic and cultural rootedness. Questions of integrity and commodification of objective authenticity cannot be dismissed when architecture of a particular building or place undergoes transformation. In many cases, as Henderson describes, "the result is a semblance of authenticity, exploited in marketing narratives which cultivate myths and are infused with nostalgia for a bygone era" (2013, pp. 452–453). The authentication process calls for a delicate compromise so that the negotiated authenticity strategy conserves the cultural soul of the heritage building, relaunched as a heritage hotel.

Authentication process

Literature directly focusing on the authentication process, in the context of heritage hotels, is nominal. However, several inferences can be drawn from existing studies. This section first scans the limited work of authors who have discussed the authentication processes in the context of heritage hotels before deriving presumptions from other studies.

In their scrutiny of an abandoned heritage hotel (Estoril Hotel) in China, Xie and Shi suggest an authenticating co-creation strategy. Four fundamental tenets of co-creation are presented: dialogue (two-way interaction), access (offering information and expertise outside of the traditional channels), risk (minimizing ambiguous communication with guests/visitors), and transparency (earning trust) (Prahalad & Ramaswamy 2004). It is postulated that authentication is a subjective phenomenon where authenticity can be appropriated by a variety of stakeholders. According to Xie and Shi, "the process of authentication offers a particularly good vantage point from which to identify specific groups and interests who make claims for legitimizing their preferred constructions of cultural identity" (2010, p. 71). Stakeholder cooperation and collaboration are critical. Additionally, different types of conservation techniques are required to breathe life and visibility into the aging or abandoned historic structures.

Table 6.2 presents a synopsis of the Estoril hotel's history and ongoing debate surrounding its future. Xie and Shi develop a theoretical paradigm to illustrate the role of authentication in the co-creation process, with the involvement of key stakeholders, aimed at developing an authentic heritage identity. The authors advocate a negotiated authentication strategy to revive the historical grandeur of the abandoned hotel. They recommend a co-creation plan to authenticate 'a preferred heritage identity' by

Table 6.2 The Estoril Hotel

Estoril Hotel was first opened in 1952 and it was initially a building with one floor. It was used as a public restaurant and a dance place. Later, three more stories were added and the building was altered into a casino hotel. A frontage of mosaic murals was then attached to the front elevation. More rooms and gambling spaces were added in 1964.
However, the casino hotel was suddenly closed in the 1990s and has remained in a state of disrepair. A number of new projects were designed for the property over the years, but they did not see fruition.
The building and its design are reminiscent of contemporary architecture and urban style from the 1960s. Perspectives on the use of the property remained fractured over the fate of the ramshackled hotel. Some opined that it showcases an important part of Macau's gambling history, hence holds historical value, while others call for an adaptive use of the building.
The Macau government has considered demolishing the hotel complex and opening a cultural and entertainment center and the proposal received the support of the majority of the local residents. However, the government's plan of razing the property has been opposed by several organizations, on grounds of its conservation and architectural value. These organizations argue that the property holds historical and emotional significance.

Source: Xie and Shi (2019).

acknowledging its past worth as a functional space and as an entertainment venue. They write:

- Co-creation serves as the nexus in debates about heritage hotel revitalization, where authentication and identity are at stake. Government, community organizations, and residents exert a huge influence in revitalizing the Hotel Estoril, which eventually leads to a strategic collaboration;
- The process of authentication contributes to a deeper understanding of co-creation methods, including various factors such as engagement platforms, trust, transparency, and interactions; and
- Stakeholders' involvement and effective information sharing methods are the direct result of authentication through co-creation. It offered an innovative path (2019, pp. 70–77).

Xie and Shi further share that some organizations "also question the vision of Macau's future when government advocates razing the hotel and replacing it with a cultural complex and what versions of authenticity get brought up in debates over the hotel's future. A survey of residents, conducted by the authors, revealed that residents agreed that the idle building was wasteful and expressed a preference for recreating the hotel at its current location. The esthetics and design of the Hotel appear to be important determinants in residents' perception of authenticity. Although esthetic perception is characterized as subjective, it has a high influence on residents' views of adaptive reuse, as a new design must reflect the identity of the hotel's history" (2019, pp. 72–75). Beyond the work of Xie and Shi, indirect inferences for heritage hotels and resorts can be extracted from other relevant literature, that offer insights on the authoritative position of authenticating agents and prejudice in the selection of markers to define/showcase objective authenticity. From this standpoint, negotiated authentication becomes a key push factor behind the innovative mechanisms, driven by a variety of reasons such as: authority and power relationships; safeguarding of heritage value and conservation of historical structures and; attracting tourism markets.

Closing comments

Negotiated authenticity is apparent in the commodification process pursued by heritage hotels; albeit on a measuring scale with high-end negotiation and low-end negotiation of the purest form of authenticity at the two ends of the spectrum, some heritage hotels and resorts are positioned at the low-end with delicate touches of adaptation. But this stance is not universal. From a sustainability standpoint, Henderson (2001) has lamented over the priority given "to making a historic property pay" and the manner in which past is exploited through strategically designed promotional strategies. Teo and

Huang (1995) argue that heritage hotels capitalizing on colonial heritage are focused on inducing nostalgic memories for a privileged past that is often distanced from the local life styles. Because many heritage hotels are located in historic locations (for example – the Culloden Hotel in Scotland), the settings themselves objectively authenticate the hotels even though the design and facilities are commodified to gratify the needs of the preferred markets. Unarguably, the innovative pursuit of authenticating heritage properties for tourism aims to accomplish a dual purpose: economic numerations and conservation. For instance, "the Temple Hotel of Japan provides dual benefits: a religious and learning environment for visitors and economic benefits for the monks" (Chhabra 2010, p. 142).

Debates associated with contested and dissonant heritage also apply to these heritage settings. As an instance, colonial narratives offer a harmonious view of the colonial era, which is antithetical to the narratives which lament that colonial imperialism was imposed on developing countries (Henderson 2013). Several takeaway points emerge from this chapter that can serve as a guide to the development of a resilient authentication process devoted to promote longevity of heritage structures. First, authentication strategies should be drawn from initiatives, which are centered on localized traditional offerings, to promote social and cultural capital. Second, balanced mechanisms are needed to support preservation of historic buildings and authentication of heritage tourism experiences, by nurturing local culture and histories. Merited authentication initiatives support negotiated authenticities of heritage hotels and resorts and advocate their patronage in a sustainable and resilient manner. Third, more work is needed to gather insights on how heritage hotels can negotiate their role as authenticating agents of tangible and intangible heritage experiences so that they remain resilient in times of disequilibrium.

Although most of the above insights were penned during the pre-pandemic times, they aim at delicate and rationalized mediations to ensure cultural continuity and penchant for essentialist/objective authenticity. Hence, negotiated endurance will continue in the post-pandemic era as adaptivity and delicate alterations are required to situate these properties in contemporary context. Evolution is a part of the human life and tourism has proved to be resilient and transformative both during and after a catastrophic phenomenon. Structural changes and cultural negotiations will continue to be rationalized as the hospitality demand is likely to shift towards more localized travel (Ioannides & Gyimóthy 2020) because of mobility restrictions (Romagosa 2020). The 'business as usual' model is unlikely to return for some time. As the pandemic continues to caste its shadow on heritage tourism, the heritage resorts are focusing on local target markets and designing delicately authenticated heritage staycation packages.

There are early signs for surge in proximity markets (Romagosa 2020) and these markets seek localized products and experiences. Therefore, traditional and negotiated versions of objectively authentic heritage will remain in demand but require ongoing negotiated authentication strategies to offer theoplacity type of experiences through self-immersions in innovatively authentic atmospheric settings. Unarguably, the pandemic will have far-reaching impact on the hospitality industry as efforts are made to identify new target markets.

With ongoing mobility restrictions and social-distancing regulations, the new normal calls for deep-rooted reformations in the manner authenticity is showcased and experienced in heritage properties. How the heritage hotels and resorts reset their authenticity and authentication strategies during post-COVID-19 times will unfold in the near future. Next chapter carries these deliberations forward, in the context of homestays. It offers insights on the popularity of this niche category of heritage accommodations. It discusses initiatives to address pandemic disruptions, especially in the developing world, and identify ways to capture the attention of new target markets particularly those confined to their backyard thresholds. The test is to transform that attention to safe close-by immersive heritage travel for authentic staycation experiences.

References

Alsisar Haveli (2019). Alsisar Haveli. Retrieved on September 5, 2019, from: https://www.alsisarhaveli.com/home.php

Aslam, M., & Joliffe, L. (2015). Repurposing colonial tea heritage through historic lodging. *Journal of Heritage Tourism*, *10*(2), 111–128.

Chang, T. (1997). Heritage as a tourism commodity: Traversing the tourist-local divide. *Singapore Journal of Tropical Geography*, *18*(1), 46–68.

Chang, T., & Teo, P. (2008). The Shophouse Hotel: Vernacular heritage in a creative City. *Urban Studies*, *46*, 341–367.

Chhabra, D. (2010). *Sustainable marketing of cultural and heritage tourism*. UK: Routledge.

Chhabra, D. (2015). Cultural hospitality framework of heritage accommodations. *Journal of Heritage Tourism*, *10*(2), 184–190.

Choo, P. W., Tan, C. L., & Yeo, S. F. (2018). A review of customer experience quality measurement in Malaysian heritage hotels. *Global Business and Management Research*, *10*(1), 379–395.

de Klumbis, D., & Munsters, W. (2005). *Developments in the hotel industry: Design meets historic properties*. Retrieved October 12, 2011, from http://bibemp2.us.es/turismo/turismonet1.

Ebbe, K. (2009). Infrastructure and heritage conservation: Opportunities for urban revitalisation and economic development. Directions in Urban Development Note. World Bank Urban Development Unit, February.

Gholitabar, S., & Costa, C. (2018). Assessing patrons' satisfaction with the cultural heritage attribute (accommodation) in the historical city Isfahan, Iran (Abbasi Hotel). *Revista Turismo & Desenvolvimento (RT&D)/Journal of Tourism & Development*, (29).

Greffe, X. (2004). Is heritage an asset or liability? *Journal of Cultural Heritage*, *5*, 307–308.

Henderson, J. (2010). Sharia-compliant Hotels. *Tourism and Hospitality Research*, *10*(3), 246–254.

Henderson, J. (2011). Hip heritage: The boutique hotel business in Singapore. *Tourism and Hospitality Research*, *11*(3), 217–223.

Henderson, J. (2013). Selling the past: Heritage hotels. *Tourism*, *61*(4), 451–454.

Henderson, J. C. (2001). Conserving colonial heritage: Raffles Hotel in Singapore. *International Journal of Heritage Studies*, *7*(1), 7–24.

Ioannides, D., & Gyimóthy, S. (2020). The COVID-19 crisis as an opportunity for escaping the unsustainable global tourism path. *Tourism Geographies*, 1–9.

Keeps, D. (2006). Historic boutique hotels. Travel and leisure. Retrieved on September 10, 2010 from: http://www.travelandleisure.com/articles/the-nextboutique-hotel-history-lesson.

Langston, C., Wong, F., Hui, E., & Shen, L. (2008). Strategic assessment of building adaptive reuse opportunities in Hong Kong. *Building and Environment*, *43*(10), 1709–1718.

Lee, W., & Chhabra, D. (2015). Heritage hotels and historic lodging: Perspectives on experiential marketing and sustainable culture. *Journal of Heritage Tourism*, *10*(2), 1–8.

Murzyn-Kupisz, M. (2013). The socio-economic impact of built heritage projects conducted by private investors. *Journal of Cultural Heritage*, *14*, 156–162.

Ozdemir, I., & Met, O. (2012). The expectations of Muslim religious customers in the lodging industry: The case of Turkey. In A. Zainal, S. Radzi, R. Hashim, C. Chik, & R. Abu (Eds.), *Current issues in hospitality and tourism research and innovation* (pp. 323–328). London.

Pongsempol, C., & Upala, P. (2017). Impacts of Adaptive Reuse of Heritage Buildings to Small Hotel Buildings in Bangkok. 5th American Conference on Quality of Life, Nouvo City, Bangkok.

Prahalad, C., & Ramaswamy, V. (2004). *The future of competition: Co-creating unique value with customers*. Boston, MA: Harvard Business School Press.

Romagosa, F. (2020). The COVID-19 crisis: Opportunities for sustainable and proximity tourism. *Tourism Geographies*, 1–5.

Rosenberg, P., & Choufany, H. M. (2009). *Spiritual Lodging – The Shariah Compliant Hotel Concept*. Retrieved August 10, 2019, from: https://www.hospitalitynet.org/opinion/4041066.html.

Saad, H., Ali, B., & Abdel-Ati, A. (2014). Sharia-compliant Hotels in Egypt: Concept and challenges. *Advances in Hospitality and Tourism Research*, *2*(1), 1–13.

Sahafa (2008). *Egypt: Hyatt and dry – Saudi hotel owner takes the fizz out of Cairo's tourist allure*. Retrieved August 2, 2019, from: http://www.theguardian.com/world/Egypt.

See, G. T., & Goh, Y. N. (2019). Tourists' intention to visit heritage hotels at George Town world heritage site. *Journal of Heritage Tourism*, *14*(1), 33–48.

Singh, K. (2006). *Rajasthan*. New Delhi: Lustre Press, Roli Books.

Teo, P., & Huang, S. (1995). Tourism and heritage conservation in Singapore. *Annals of Tourism Research*, *22*(3), 589–615.

Timothy, D., & Teye, V. (2009). *Tourism and the lodging sector*. Oxford: Butterworth Heinemann.

Tollotson, G. (2011). *Umaid Bhawan Palace*. Jodhpur: Meharangarh Museum Trust.

Upadhyaya, V. (2017). Transformation in traditional havelis: A case of Walled City Jaipur, *Rajasthan Imperial Journal of Interdisciplinary Research*, *3*(2), 1482–1491.

Xie, P., & Shi, W. (2019). Authenticating a heritage hotel: Co-creating a new identity. *Journal of Heritage Tourism*, *14*(1), 67–80.

7 Negotiated authenticity and authentication of homestay tourism

This chapter discusses the authenticity of homestays. Their potential as a form of sustainable tourism is also critically examined. The chapter closes with offering insights on the survival of homestays and their patronage during the intra- and post-pandemic times.

During the past several decades, homestays have become an important asset of community-based tourism, especially in the rural regions of the globe. These small micro enterprises promote sustainable consumption of cultural resources, generate socioeconomic benefits for the host community, and contribute toward preservation of the natural environment (Ellis 2000; Jamal, Othman, & Muhammad 2011; Lynch, Di Domenico, & Sweeney 2007). Furthermore, the homestays have shown potential to reset the disequilibrium between the privileged and poor communities, especially in the remote and/or fragile regions of the developing world. The homestay programs have improved local quality of life by promoting livelihood (SL) opportunities (Agyeiwaah, Akyeampong, & Amenumey 2013; Ahmad, Jabeen, & Khan 2014; Hussin, Kunjuraman, & Weirowski 2015; Ibrahim & Razzaq 2010; Jamal et al. 2011; Lynch 2005; Pakshir & Nair 2011; Rasoolimanesh, Dahalan, & Jaafar 2016; Sood, Chhabra, & Andereck 2017).

Most homestay programs receive support, from the local and state governments, because they serve as conduits for the sustenance of local culture and indigenous traditions (Rasoolimanesh et al. 2016). For instance, the homestay program in Malaysia "aims to facilitate community participation to stimulate economic benefits and offer tourists a way to experience traditional life in a Malaysian village" (Ahmad et al. 2014, p. 74). Furthermore, interactive participation of hosts, in sustainable tourism programs, can amplify livelihood opportunities by selling authentic local handicrafts (Hussin 2002; Pakshir & Nair 2011). Such initiatives can breathe life into declining handicraft industries and make them flourish, while at the same time inculcate a sense of pride in local traditions (Hussin et al. 2015). Hussin et al. recognize the significance of strategic planning to enshrine the authenticity of a homestay program while making it economically viable (Cohen 1985). As an alternative form of accommodation that takes place in rural and often remote communities, their authenticity and resilience is worthy of exploration. This chapter discusses the authenticity of homestays. Their potential as a form of sustainable tourism is also critically examined. Finally, the chapter closes with insights on their survival strategies and patronage in the pandemic and post-pandemic times.

Authenticity of homestays

The homestay concept, at its core, is premised on the objective authenticity notion because its experience is rooted in local and cultural/traditional attributes. This can be evidenced in the manner homestays have been defined in extant literature.

Lynch refers to homestays as "types of accommodation where tourists or guests pay to stay in private homes, where interaction takes place with a host and/or family usually living upon the premises, and with whom public space is, to a degree, 'shared' (2005, p. 528). According to Jamal et al. (2011), homestay programs permit a match between tourists and selected local families and offer an opportunity to experience a part of their daily life. Hjulmand et al. (2003) say that, "the idea of homestay is to accommodate tourists in a village with a local family, thus enabling the tourist to learn about local lifestyle, culture, nature". Different local settings are used for homestays across the globe. For instance, farmhouse accommodations are promoted as homestays in Australia. According to Tourism Malaysia, they offer opportunities to tourists to immerse in different cultures and acquire new knowledge. Table 7.1 offers examples of different types of homestay programs across the globe.

Evidently from the foregoing, homestay programs are rooted in different types of local traditions. From a marketing standpoint, several studies have examined visitor perceptions and experiences of homestay tourism so that

Table 7.1 Different types of homestays

Country	*Category*
Canada	Cultural/Heritage and Farmstay
USA	Farmstay/Agricultural and Educational
Australia	Farmstay
Japan	Home Experience and Educational
Malaysia	Village-style Living and Heritage
India	Heritage/Cultural
South Korea	Educational
Thailand	Educational, Volunteer Work, and Cultural
Gautamela	Educational and Cultural
Spain	Educational/Language
Singapore	Educational
Indonesia	Cultural and Leisure
United Kingdom	Family and Educational
New Zealand	Educational, Cultural, and Farmstay
Switzerland	Family, Natural, and Educational

Source: Leh and Hamzah (2012, p. 7608)

sustainable promotional strategies, premised on authentic experiences, can be recommended to the local government and the homestay communities (Rasoolimanesh et al. 2016; Sood et al. 2017a). To date, marketing-related research on homestays has focused predominantly on tourist motivations, quality of stay (often used as a proxy for objective authenticity), perceptions, satisfaction, and loyalty. For instance, the homestay program in Malaysia is promoted for its unique experiential setting with the host families and local communities. The guests are offered opportunities to participate and immerse themselves in host culture through interactions and engaging activities such as cooking and handicraft making (Kalsom 2009). A village homestay program is operated by a group of certified homestay operators to ensure quality is retained.

In his study on homestays in Sri Lanka, Ranasinghe et al. (2015) uses motivation as an endogenous construct while exogenous factors used are satisfaction, quality perceptions, and loyalty. The results suggest a strong relationship between motivation for authentic objects and experiences and satisfaction among homestay tourists. Authenticity influences satisfaction as evidenced from the significant relationship reported between several motivational items associated with authenticity (namely local culture, family togetherness, social interaction with the hosts, and educational experience) and satisfaction. Jamal et al. (2011) examine different dimensions of perceived value in the context of community-based homestay tourism (CBHT). CBHT is defined as "a form of tourism that is closely related to nature, culture and local custom and is intended to attract a certain segment of the tourist market that desires tourist experiences" (Jamal et al. 2011, p. 5). Their results report a significant relationship between perceived value (especially in the context of authenticity of cultural experience) of homestay experience and satisfaction. Perceived value dimensions include "experiential value of guest-host interactions, experiential value of activities, culture, and knowledge, emotional value and functional value" (Jamal et al. 2011, p. 57). The study argues that homestay visitors are not homogeneous albeit many seek immersion in 'othered' experiences. Rasoolimanesh et al. (2016) also examine the link between homestay tourists' perceived value and satisfaction and confirm positive and significant influence of perceived value on satisfaction.

Homestay – a form of sustainable tourism

Clearly, homestay tourism is premised on several (if not all) sustainability benchmarks. Extant literature presents it as a subset of rural tourism, pro-poor tourism, indigenous tourism, ecotourism, heritage tourism, or volunteer tourism. Sood et al. (2017b) postulate that the key selling point of a homestay is its rich cultural heritage and family traditions. Local community is strongly committed to the promotion of local custom and traditional cultural norms. Subsequently, the pro-poor tourism development strategy is motivated by both economic and non-economic goals. Naipinit and Maneenetr (2010) examine benefits of community participation in the context of homestay tourism and identify issues impeding this initiative. The biggest problem is related to inadequate public utilities followed by lack of respect/pride in the local culture and norms, lack of support from the local organizations, and sparse opportunities to earn supplementary income in the village. To address these issues, numerous suggestions are offered by the authors (presented in Table 7.2).

From the aforementioned, it is clear that equitable representation and better coordination with the local government and other stakeholders are required to sustain long-term victory of homestays. An authentic experience and a pristine/unique setting solely cannot make a homestay program successful; equally important is the infrastructure and support from other role players in the tourism system. While offering a discourse on rethinking sustainable tourism, Khamdevi and Bott (2018) share positive outcomes of offering rural homestay programs, using Bali (Indonesia) as a backdrop. Mass tourism had depleted the natural environment in Bali and compromised its cultural authenticity. Khamdevi and Bott (2018) argue that rural

Table 7.2 Enhancing community participation in management of homestay tourism

No.	*Suggestions*
1	The community should cooperate with local organizations to discover the best tourism management policies for the village; however, the villager participation needs to be facilitated by the local organizations by offering the community a platform to voice its views and share opportunities for involvement in local tourism events.
2	Meetings are rarely attended by the locals. It is suggested that more efforts are required to educate the locals and convince them to get involved and attend meetings to help plan responsible tourism management.
3	Equitable participation across different communities in the village is crucial to brainstorm practical ways to clarify policies and avoid conflicts.
4	Though the level of participation associated with topics such as income and investment is found to be robust, all stakeholders and local residents need to support village businesses, as a community enterprise.
5	The level of participation related to talks on investment in planning operations is high, but the village needs public sector support to develop public utilities such as water supply, electricity, and roads. The village communities need to cooperate with the government, so that utilities can be planned and offered in an efficient manner.
6	The village continues to have a low level of participation in areas focused on the assessment of restoration and the construction of attractions. It is important for the entire village community to solicit assessment of relevant agencies to demonstrate consensus and support.
7	The village continues to receive minimal information directly from the government. To resolve this issue, the government should creatively figure out ways to disseminate important and detailed information to the village.

Source: Naipinit and Maneenetr (2010)

tourism programs offer an alternative form of sustainable tourism and these can be based on authentic rural experiences/activities in local farms and other rural settings. Alternative and viable forms of tourism prospects can enhance the cultural veracity of local communities and help inculcate a sense of character, authenticity, tenacity, and accomplishment. As the authors illustrate sustainable tourism offers opportunities to learn and immerse in the lives of other cultures and "tourists can contribute to preservation of the tangible and intangible heritage (Khamdevi & Boycott 2018, p. 8). Table 7.3 offers an inside overview of initiatives taken to introduce homestay tourism in Malaysia. It maps its history and shows how objectively authentic elements are sustained through support garnered from different stakeholders. Malaysia is an insightful example (such as offering hands on experience with handicraft making) of how a complete heritage tourism package is designed for homestay tourists who seek an immersive experience.

Documented literature also notes that several Indian homestay programs offer ideal settings to promote sustainable consumption of cultural and rural resources and meet tourist's quest for an authentic, novel, and personalized experience (Lynch et al. 2007; Sood 2016). According to Kontogeorgopoulos, Churyen, and Duangsaeng, these "rural homestays allow guests to catch a glimpse of the daily lives of village residents and therefore serve as a means of experiencing a local community in ways that differ from conventional tourism interactions and settings. As an alternative

Table 7.3 Homestay as a pro-poor tourism strategy in Banghuris (Malaysia)

- The Banghuris Homestay is a popular homestay in Sepang (Selangor State), Malaysia, and is known for its traditional village hospitality. The Banghuris name represents three villages (Kampung Bukit Bangkong, Kampung Hulu Chuchuh, and Kampung Hulu Teris). It is 30 kilometers from the Kuala Lumpur International Airport and houses more than 150 guest rooms.
- The idea of introducing the Baghuris Homestay took shape when the Kampung Hulu Chucoh village received the Selangor State Best Village Competition award in 1992. This award gave visibility and started attracting many visitors. The homestay program received the support of various agencies such as the Sepang District Municipal Council, the Ministry of Tourism (previously called the Ministry of Culture, Art and Tourism) and numerous travel agencies. The program started with 18 houses and then it expanded to 68 houses.
- The Banghuris Homestay is the only homestay in the Sepang District with Javanese Maly background. Popularly sought activities in Sepang include agro-tourism activities such as a visit to the palm plantations, rubber plantations, coffee farms, fruit orchards, and small micro enterprises. A range of authentic local folk activities have also been transformed into tourist attractions. Tourists also enjoy visits to the historical Bukut Bangkong, and like to play traditional games with the villagers, and handicraft making.
- The Banghuris Homestay has won many accolades and was the winner of Zon Tengah level Ilham Desa Competition (in 2005). It also received the Ministry of Tourism award in addition to the Malaysia's Best Homestay Award (in 2004).

Source: Leh and Hamzah (2012).

form of accommodation that features basic standards and takes place in small, and often remote, rural communities, homestays appeal to a very small and specific niche of tourists" (2015, p. 31). Additionally, most of the remote mountain areas are not easily accessed due to poor infrastructure. From a positive standpoint, this lack of access shielded them from commodification and their traditional cultures have remained authentic.

In fact, in India, the idea behind rural tourism development (of which homestays are a crucial part) has been to offer authentic cultural experiences to tourists by giving them an opportunity to stay with rural communities and witness their livestyles, traditions, and ways from proximity. This new shift in focus from a commercial hotel-centric vision to a rural experience-centric approach developed with the launch of the Endogenous Tourism Project-Rural Tourism Scheme (ETP-RTS) in 2003. It was the initiative of the Ministry of Tourism (MoT), Government of India, in collaboration with United Nations Development Program (UNDP). On similar lines, many Himalayan states including Himachal Pradesh launched their own Homestay Schemes. For centuries, the Himalayas have been a getaway to a subliminal and transcendental world. In the fragile Himalayan ecosystems, homestays are an opportunity to render existing built structures (houses) for sustainable cultural consumption and facilitate community involvement, preservation of ecology, and authentic cultural heritage.

Under the Himachal Homestay Scheme (2008), a local resident can rent up to three rooms of his house to tourists; but the house has to conform to certain guidelines. Incentives such as subsidies on electricity and water and tax exemption on income from homestays are offered to expand participation. Policymakers are vigorously promoting homestays, in remote Indian Himalayan regions, as an authentic pathway to stimulate rural economy, create employment opportunities, and prevent migration of local youth to urban areas (Sood, Lynch, & Anastasiadou 2017). The next few paragraphs share an overview of the history of authentic homestay programs in the State of Himachal Pradesh. Insights are shared about the manner in which cultural richness of remote regions is experienced in local homes. These homes serve as living museums in remote rural settings and their demand has given impetus to the success of lifestyle rural entrepreneurs.

History of authentic homestay development in Himachal Pradesh

Homestays have been a part of tourism development, since the nascent stage of tourism, in the State. Local communities began hosting tourists in their houses, since the 1980s, due to very few hotels and guest houses in Kullu (a popular Himalayan tourist destination). At that time there were no travel agents or telephones in the Himalayan hills stations of Kullu, Shimla, and Dalhousie. When the tourists arrived, a guide or a broker guided them to local accommodations. A broker, according to Cohen (1985), is a person

Figure 7.1 An authentic homestay.

who is a link between the visitors and local community. In Manali, when there were no rooms available in the hotels during peak season, these guides or their friends provided meals and shelter for the night in their houses. Later on, many foreign tourists started staying in local homes, outside the villages, or situated in secluded orchards. These eventually became faces of the first form of homestays in Kullu.

These secluded 'second homes' (Figure 7.1) were spare dwellings/cottages or storage places generally in orchards or farms and were utilized by the family only during harvesting season. They became popular, especially among foreign tourists, who wanted to escape from their mundane daily life and sought out-of-the ordinary and isolated settings. They are also of interest to people who are looking for solitude and meditation in the Himalayas. Just like any other second home, these mountain cottages or houses are visited only during summer months and range from simple wood and stone structures to plush wooden lodges. They are mostly popular in the town of Kullu, although some of them are also located in villages around the Great Himalayan National Park. Tourists, to many of these homestays, have formed long lasting ties with their local hosts and have patronized them every year (Sood, 2014).

The tourists in these homestays are given unique opportunities to immerse themselves, enjoy the simple life of a Kullu household and take a glimpse of their daily activities and authentic traditions (see Figures 7.2 and 7.3) (Sood et al. 2017b). Sood et al. scrutinized the authenticity of these

Figure 7.2 Authenticating homestay experience.

homestay accommodations using multiple criteria: (1) the manner in which local homes are modified based on the homestay scheme; (2) the manner in which homes are showcased and offered access; (3) code of ethics and behavior expected of guests; (4) management of homestays (for instance – how are the sociocultural settings designed for guests); (5) manner in which the home is shared with guests; (6) the manner in which guests are informed of local history and culture; (7) type of activities offered inside and outside the home; (8) educating guests of locally produced authentic products

Figure 7.3 Authenticating homestay experience.

and souvenirs; and (9) themes used to promote the homestay. Based on the results, three forms of authenticities are pinpointed (2017):

> *Essentialist authenticity*: includes the most authentic cultural experience and hospitality is offered 'as-is;' There is little or no modification to the house, family generally stays on premises and tourists get the opportunity to live like locals.
>
> *Negotiated authenticity*: is offered by locals who have a prior hospitality experience and believe homestay should include 'hotel like' amenities. Tourists may be provided cultural experiences on demand but within luxuries of a customized setting and so, in this context, the offering is of a blend of old and new architecture and showcase of customs in a modified setting.
>
> *Constructivist authenticity*: homestay is modified to look like a hotel or resort with all modern amenities. The family is not present on premises and the property is generally operated by non-locals purely for commercial reasons and tourists are not able to directly experience authentic culture. Table 7.4 presents a breakdown of items that resonate with each dimension of authenticity.

The most traditional homestays are threatened by the proliferation of commercial resorts, hotels, and guest houses. In competition with these commercial accommodations, more and more homestays have imitated the services offered by hotels. This has compromised their traditional authenticity. Sood et al. writes that:

> Homestays in Kullu range from aesthetically beautiful alpine wooden cottages to simple wood and stone houses with traditional Kulluvi architecture. These homestays fulfill a tourist's desire to live like a villager. These are mainly run by the local people and offer village life experiences 'as- is' (essentialist).
>
> However, these traditional homestays only capture a fraction of the total tourists to the region. Homestay hosts have been vying to woo customers by lowering their prices and trying to emulate hotels in services and facilities. Constructivist authenticity is offered in some homestays where homestays are commodified to meet the market demand. Sharing platforms like Airbnb are being exploited by hoteliers to camouflage their properties as homestays and tourists feel cheated. Eco tourists, who select to live in homestays, are disappointed to miss out on the fundamental promise of a home stay: the authentic Himalayan lifestyle experience. (2017a, pp. 3–5)

When locals open their homes for tourists, they select certain aspects of their life and culture and showcase them in the shared spaces. During the

Table 7.4 Types of authenticity observed in homestays in Indian Himalayas

Type of authenticity	*Categories used*	*Homestay activities*
Essentialist	• Represents the past heritage • True to the original version • Old and ancient	Tourists are served food in a traditional setting Tourists learn how to make local cuisine from the homestay host Local cuisine such as *Siddu* is offered to the guests If tourists demand, some dance or music performance is organized from local village groups Décor and architecture of the house, dining, and living areas is original which reflects the local architecture
Negotiated	• Offers interactive opportunities with locals • Offers opportunities to participate in traditional/folk activities in an adjusting/flexible manner	Certain areas of the house are 'off-limits' to the guests Artifacts and household items, showcased in the homestay, reflect the local ambience Some signage is posted to display rules of the house e.g. guests are requested to leave their shoes outside the house Kitchen design is a mix of traditional and modern architecture Internet facilities are provided in even remotely situated homestays
Constructivist	• Commodified to meet market demand • Staged • Managed and regulated	Some homestays copy hotels to offer stylish bedrooms and baths Only a handful of homestays use the word 'hotel' or 'guest house' in their names Toilets are customized based on tourist needs and not according to local design and architecture in some homestays
	• Reconstructed to embrace modern culture	Restaurant like setting is provided in a homestay for to facilitate modern dining experience. One rural homestay even offered a bar Some homestays have added hotel-like amenities such as table tennis, laundry, cable TV, parking, room service, complementary toiletries, and tea maker

Source: Sood et al. (2017)

stay, access to private spaces of the house is often blocked, with signs to remind the tourists to stay away. Hence tourists get a glimpse of selected showcasing in a controlled setting (Thapa & Malini 2017). This can be termed as staged authenticity. However, at the same time, it is also noted that local hosts are keen to showcase their lifestyles and offer engaging experiences, in an objectively authentic manner (Table 7.4). At the same time, they are compelled to protect their backstage, so some parts of the home are made inaccessible and guest movement is restricted to the living room and the guest rooms. An informal code of expected ethical behavior is often communicated to facilitate visitor mindfulness. To some extent, tangible evidence of object authenticity is visible to satisfy the tourist desire for a genuine cultural experience. For instance, guests are invited to participate in traditional activities in staged performative spaces.

Competition from accommodations, owned by non-locals, has resulted in modifications and compromise of host traditions (resulting in constructivist authenticity-type of experiences). Many times, the mediating agencies fail to direct tourists, seeking genuine rural experiences, to authentic homestays. In some cases, poor marketing efforts fail to communicate/promote the benefits of staying in a homestay, that is, in terms of an authentic cultural experience and an actual glimpse of rural Himalayan life. To some extent, interview results of Sood's study reveal that local homestays are also commodifying themselves by promising 'hotel-like' services and popular/contemporary cuisine (2014). Hosts are stretched between the desire to present an 'original' rustic (objectively authentic) experience and 'comfortable hotel-like' hospitality experience. The outcome is commodification (or selective negotiation), in the effort to offer customized experiences. The newly opened homestays, particularly, are vulnerable to commodification, external forces, and competition triggered by the explosion of external agents (Sood et al. 2017b).

One might question if and to what extent these objectively authentic homestay programs are able to impact the socioeconomic status of a local community, particularly in terms of new income (Leh & Hamzah 2012) and pride in the local traditions. To give insights on the impact of planning and tourists, a micro-level view from a rural village (Korzok) is offered here. The purpose of this narrative is also to illustrate that authenticity is just one piece of the sustainable development initiative, and several dimensions/issues have to be considered to strategize successful SL initiatives (Anand, Chandan, & Singh 2012). Backed by NGOs like ECOSS, Mountain Shepherd, Ecosphere, UNESCO, and State and Central governments, CBHT has been launched in Korzok. The topography of the harsh region, due to high elevation and oxygen levels (such as Ladakh in the northernmost region of India) poses serious challenges in terms of infrastructure and operational constraints. Anand et al. posit that, although authentic homestay tourism is a good supplementary source of income, it needs to be tapped in a strategic and sustainable manner.

Table 7.5 Homestays in Korzok

- The climate is subarctic with temperatures ranging from 30 degrees centigrade in summer to –40 degrees centigrade in winter.
- This village is one of the very few elevated settlements in the world at an altitude of approximately 15,000 feet above sea level.
- The nearest urban center (Leh) is 215 kilometers away.
- It is connected by a single-track road that becomes inaccessible in winter.
- Population is 1300 households out of which 22 households live in permanent structures but they also migrate seasonally; the rest of the population consists of nomads who live in tents, made of Yak hair or skin.
- There is one primary health center in the village.
- There used to be one primary school, but now all students have been shifted to a centralized school in the Puga Valley.
- A diesel generator was provided by the local hill council that provides electricity to the local village for 4 hours in the evening.
- Seventy percent practice subsistence agriculture. The cultivation season is short and most pastoralists are nomads.
- Barley, wheat, and peas are grown mainly for fodder.

Source: Anand et al. (2012)

Korzok forms a unique setting of scrutiny in terms of its location, authentic culture, socioeconomic characteristics and history. World Wide Fund (WWF) India spear headed the homestay initiative in 1999 to facilitate economic development and pro-poor tourism. The homestay model in Korzok (Ladakh) was developed by the Snow Leopard Conservancy with a sole purpose to balance development and cultural/natural conservation initiatives. Table 7.5 presents an overview of Korzok village.

The indigenous Changpas, in Karzok village, live in a harsh cold desert environment. They are pastoralists and their livelihood is deeply embedded in natural resources. They practice subsistence agriculture on marginal lands and raise pashmina goats or Changra goats, their main occupation for three to four centuries. A territorial dispute between India and Pakistan has challenged their natural and cultural resources. It has resulted in a large and lasting military presence that has catalyzed change from the traditional subsistence economy and trade to dependence upon a market economy and subsidized goods imported by the central government (Goodall 2004); they are seeking alternative forms of employment to supplement their incomes, although they do not necessarily possess skills or training to make these transitions. Anand et al. further report that the key issue faced by this remote community is lack of expertise in developing strategies that are in harmony with the natural environment. According to Wunder (2000), the economic perspectives of conservation, in remote regions, require crucial attention, because of the harsh environment conditions. The Changpas have received funding support from the WWF to strategically plan tourism development in a way that conserves the local traditions and does not damage the high-altitude wetlands. In fact, the WWF has made several initiatives to study the feasibility of homestay programs in the region. These include:

- The cultural disposition of the Ladakhi people to host visitors and the convenience of running a homestay without major investments generated interest in the initiative;
- Preliminary consultations with the community leaders and survey of all permanently settled (23) households;
- Tourism surveys to gather insights on homestay potential, desired experiences, and budgetary constraints;
- Understanding the gap in required skills, WWF-India facilitated a series of capacity-building exercises for the Changpa community with homestay facilities; and
- Capacity-building exercises.

A way forward is recommended that focuses on a "comprehensive livelihood assessment through household surveys, along with an in-depth analysis of how the local site conditions have changed over time in terms of social, economic and environmental aspects" (Anand et al. 2012, p. 135). Several insights are gathered from the green initiatives project and recommendations include:

- Members of the local population must have awareness, information, and opportunities to participate in order to make decisions about their livelihoods. This initiative gives empowerment to the Changpa;
- Equitable access to natural resources and their use and responsible management by various stakeholders is essential to minimize conflict;
- Social and economic incentives are a must for community involvement in the conservation process;
- Communities have to be empowered through financial and technical assistance, and skill-building exercises for a fair and smooth transition to adopt alternative sources of income base on pro-environment policies;
- Cultural competence and sensitivity are required for aligning local livelihoods with the broader goals of poverty reduction, development, and conservation;
- Local cultural and spiritual value systems should be represented and integrated in the projects. Because the communities have stayed reslient in these areas for centuries, they are spontaneously disposed to support natural conservation activities in the area; and
- Reinforcement of compliance protocol, requires a certification system, across the region and at various operational layers to facilitate green economy projects; initial stakeholder views were ascertained for developing the certification system (Anand et al. 2012).

Closing comments

As illustrated in the Korzok example, ecological disturbance and extreme commodification (constructivist version) of authenticity have posed a threat to traditional homestay programs. Pristine regions can be contaminated if development of commercial accommodations, such as Airbnbs and hotels owned by outside corporations, is not curbed. Homestays are a viable sustainable fix as they promote conservation of local heritage and hold economic potential. Sood et al. postulate that, "local policymakers and UNESCO need to step up their efforts to help local hosts to develop objectively authentic homestay experiences and check the development of non-locals' owned accommodations. Careful blending of existentialist and objective authenticities is required as demand for authentic experiential consumption continues to emerge" (2017a). The most meaningful authenticating agents are the homestay hosts who, with appropriate education and incentives, can perform the role of *lifestyle entrepreneurs* to showcase and strategically orchestrate negotiated authentic homestay experiences. As entrepreneurs, they can strive to creatively compete with commercial accommodations and leverage their authentic settings and traditions to sell novel cultural experiences.

According to McIntosh et al., "the commercial home concept has significant emotional importance to its owners and the impact of hosting on the hosts' personal and family life, and private space, can be immense" (2010, p. 3). The perspective of hosts and the emotional relationship with the home constitute an important field of humanistic inquiry, especially in the context of vulnerability and resilience theories, to gather insights on how they can sustain their cultural norms and authentic way of life while selectively sharing them with tourists. Scrutiny is required to figure out resolution of challenges faced by lifestyle entrepreneurs, in the wake of encroachment by external corporations, offering accommodations under the 'homestay' umbrella. By the same token, it is also important to understand the cultural and social context in which local lifestyle entrepreneurs negotiate, open, and showcase their home to outsiders. A lifestyle entrepreneur's vulnerability is his or her ability to anticipate, compromise, cope, and recover from shocks over an extended period of time and is dependent on access and claim to related sociopolitical, economic, and environmental resources and power systems (Calgaro, Lloyd, & Dominey-Howes 2014).

According to Calgaro et al., "human agency also plays a critical role in influencing differential vulnerability patterns. Understanding why actors choose certain actions over others requires a deeper knowledge of people's risk perceptions, assumptions, past experiences, personal characteristics and values systems, culture and economic circumstances. Resilience refers to the capacity of a system to absorb disturbance and reorganize throughout periods of change, while retaining authentic function, structure and identity" (2014, p. 343). It is argued that "impact factors and coping responses to shocks depend on the set of available capital, including the effectiveness of governance structures, levels of preparedness, and capacity to learn at the time of the shock's impact or breaching of the stressor's tipping point (Tompkins & Adger 2004, p. 350).

Authentic homestay tourism holds tremendous potential to strategically promote negotiated authenticity and the livelihood of lifestyle entrepreneurs. Evidenced in the parallel tenets of pro-poor tourism, ecotourism, community tourism, sustainable tourism, and responsible tourism (Chhabra & Phillips 2009; Ellis 2000; Liu 2006), the overall aim of homestay tourism is to promote SL of rural communities by helping them to showcase their uniqueness and authentic lifestyles. SL is a way of thinking that calls for proactive and integrative (cross-sectoral) people-centered perspectives for rural authentication of tourism, especially in remote parts of the globe (Shen, Hughey, & Simmons 2008; Su, Wall, & Xu 2016). SL approaches are participatory, people-centered, collaborative, and based on evolutionary ways of strategizing poverty reduction by developing synergies between other stakeholders and the public sector (Shen et al. 2008; Tao & Wall 2009). Key objective is to authenticate local development/management of heritage in a holistic and dynamic manner by bridging gaps between stakeholders and mitigating external influences.

In summary, homestay tourism is a SL opportunity with its unique particularities and its authenticity can be strategically negotiated by the hosts, by delicately harmonizing supply (production) and demand (consumption) perspectives (Hussein 2002; Iorio & Corsale 2010; Lee 2008; Shen et al. 2008; Simpson 2007). If not strategically devised, negotiated authenticity can lead to the encroachment of cultural traditions. Some parameters need to be observed to remain within the acceptable threshold of objective authenticity. In other words, delicate authentication processes are required so that tourism does not damage the core essence of host culture.

It cannot be denied that COVID-19 will open new pathways for sustainable tourism, although some activities, deemed obsolete and intrusive to the emerging consciousness of empathy and solidarity, will be halted (Lew, Cheer, Haywood, Brouder, & Salazar 2020). Cautiousness is evident on both sides, the guests and the hosts. While some homestay owners in the Himalayas in India have already hosted guests, as the pandemic continues to rage across the globe, others remain indecisive about opening their doors for tourists (Jangra & Kaushik 2020). Homestay tourism, especially in rural towns of Southeast Asia, has been heavily reliant on international tourists. Long distance travel has paused and this has disrupted the economy of these rural towns. On a positive note, a crisis of such a magnitude is offering opportunities to reinvent and envision an image of a truly sustainable world. Homestay tourism is likely to be recognized as one of the first frontiers of transformative tourism, once the pandemic is under control. This can be attributed to the fact that it happens at a slow pace and at isolated levels. Furthermore, localized economies are noted to be more resilient because they rely on local supply systems instead of global supply chains that have been devastated enormously by COVID-19 (Gössling Scott, & Hall 2020).

Several sustainability frameworks have been proposed by documented literature that serves as important tool to decipher the relationship between pro-poor tourism, cross-sector development, and community relations (Iorio & Corsale 2010; Su et al. 2016; Tao & Wall 2009). Although numerous SL paradigms have aimed to examine pro-poor tourism (Lee 2008; Simpson 2007; Su et al. 2016). It is argued that the conceptual underpinnings of SL programs can be extended to authentic cultural/heritage and ecotourism settings in remote or isolated indigenous communities to counter the misuse of authentic heritage resources by outside agencies (Chhabra 2010; Derrida 2000; Ellis 2000). Moving forward, in the next chapter, I extend the negotiated authenticity and its authentication dialogue to another unique expression where culture and heritage are tangibilized to brand preferred national images.

References

Agyeiwaah, E., Akyeampong, O., & Amenumey, E. K. (2013). International tourists' motivations to choose homestay: Do their socio-demographics have any influence? *Tourism and Hospitality Research*, *13*(1), 16–26.

Ahmad, S. Z., Jabeen, F., & Khan, M. (2014). Entrepreneurs choice in business venture: Motivations for choosing home-stay accommodation businesses in peninsular Malaysia. *International Journal of Hospitality Management*, *36*, 31–40.

Anand, A., Chandan, P., & Singh, R. B. (2012). Homestays at Korzok: Supplementing rural livelihoods and supporting green tourism in the Indian Himalayas. *Mountain Research and Development*, *32*(2), 126–136.

Calgaro, E., Dominey-Howes, D., & Lloyd, K. (2014). Application of the Destination Sustainability Framework to explore the drivers of vulnerability and resilience in Thailand following the 2004 Indian Ocean Tsunami. Journal of Sustainable Tourism, 22(3), 361–383.

Chhabra, D. (2010). Host community attitudes toward tourism development: The triggered tourism life cycle perspective. *Tourism Analysis*, *15*(4), 471–484.

Chhabra, D., & Phillips, R. (2009). Tourism-based development. In R. Phillips and R. Pittman (Eds.), *Introduction to community development* (pp. 236–248). London: Routledge/Taylor & Francis Group.

Cohen, E. (1985). The tourist guide: The origins, structure and dynamics of a role. *Annals of Tourism Research*, *12*(1), 5–29.

Derrida, J. (2000). Hospitality. *Angelaki*, *5*(3), 3–18.

Ellis, F. (2000). *Rural livelihoods and diversity in developing countries*. New York, NY: Oxford.

Goodall, S. (2004). Changpa nomadic pastoralists: Differing responses to change in Ladakh, North-West India. Nomadic Peoples, 8(2), 191–199.

Gössling, S., Scott, D., & Hall, C. M. (2020). Pandemics, tourism and global change: A rapid assessment of COVID-19. *Journal of Sustainable Tourism*, 1–20.

Hjulmand, L. G., Nielsen, U., Vesterløkke, P., Busk, R. J., & Erichsen, E. (2003). Tourism as a development strategy in rural areas adjacent to the Croker Range

National Park, Sabah, Malaysia. *ASEAN Review of Biodiversity and Environmental Conservation (ARBEC)*, 1–19.

Hussein, K. (2002). Livelihoods approaches compared: A multi-agency review of current practice. In *Overseas development institute, prepared for DFID*. London, UK.

Hussin, R., Kunjuraman, V., & Weirowski, F. (2015). Work transformation from fisherman to homestay tourism entrepreneur: A study in Mantanani Island Kota Belud, Sabah, East Malaysia. *Jurnal Kemanusiaan*, *13*(1).

Ibrahim, Y., & Razzaq, A. R. A. (2010). Homestay program and rural community development in Malaysia. *Journal of Ritsumeikan Social Sciences and Humanities*, *2*(3), 7–24.

Iorio, M., & Corsale, A. (2010). Rural tourism and livelihood strategies in Romania. *Journal of Rural Studies*, *26*, 152–162.

Jamal, S. A., Othman, N. A., & Muhammad, N. M. N. (2011). Tourist perceived value in a community-based homestay visit: An investigation into the functional and experiential aspect of value. *Journal of Vacation Marketing*, *17*(1), 5–15.

Jangra, R., & Kaushik, S. P. (2020). Understanding tribal community's perception toward tourism impacts: the case of emerging destinations in western Himalaya, Kinnaur. Asian Geographer, 1–24.

Kalsom, K. (2009) Community based toursim in developing countries. *Proceeding of International Seminar on Community Based Tourism*.

Khamdevi, M., & Bott, H. (2018, March). Rethinking tourism: Bali's failure. In *IOP Conference Series: Earth and Environmental Science* (Vol. 126, No. 1, p. 012171). IOP Publishing.

Lee, M. (2008). Tourism and sustainable livelihoods: The case of Taiwan. *Third World Quarterly*, *29*(5), 961–978.

Leh, F. C., & Hamzah, M. R. (2012). Homestay tourism and pro-poor tourism strategy in Banghuris Selangor, Malaysia. *Elixir International Journal*, *45*, 7602–7610.

Lew, A. A., Cheer, J. M., Haywood, M., Brouder, P., & Salazar, N. B. (2020). Visions of travel and tourism after the global COVID-19 transformation of 2020. *Tourism Geographies*, 1–12.

Liu, A. (2006). Tourism in rural areas: Kedah, Malaysia. *Tourism Management*, *27*, 878–889.

Lynch, P., Di Domenico, M., & Sweeney, M. (2007). Resident hosts and mobile strangers: Temporary exchanges within the topography of the commercial home. *Mobilizing Hospitality: The Ethics of Social Relations in a Mobile World*, 121–144.

Lynch, P. A. (2005). Sociological impressionism in a hospitality context. *Annals of Tourism Research*, *32*(3), 527–548.

Naipinit, A., & Maneenetr, T. (2010). Community participation in tourism management in Busai village homestay, Wangnamkheo District, Nakhon Ratchasima Province, Thailand. *The International Business & Economics Research Journal*, *9*(1), 103–103.

Pakshir, L., & Nair, V. (2011). Sustainability of homestay as a form of Community-based Tourism (CBT): A case study of the rural community in Bavanat-Iran. *TEAM Journal of Hospitality and Tourism*, *8*(1), 5–18.

Ranasinghe, R. U. W. A. N. (2015). Modeling visitor perceptions on homestay tourism in Sri Lanka. *South Asian Journal of Tourism and Heritage*, *8*, 18–29.

Rasoolimanesh, S. M., Dahalan, N., & Jaafar, M. (2016). Tourists' perceived value and satisfaction in a community-based homestay in the Lenggong Valley World Heritage Site. *Journal of Hospitality and Tourism Management*, *26*, 72–81.

Shen, F., Hughey, K., & Simmons, D. (2008). Connecting the sustainable livelihoods approach and tourism: A review of literature. *Journal of Hospitality and Tourism Management*, *15*, 19–31.
Simpson, M. (2007). An integrated approach to assess the impacts of tourism on community development and sustainable livelihoods. *Community Development Journal*; doi: 10.1093/cd/bsm048.
Sood, J. (2014). *Potential of homestay tourism in Himachal Pradesh: A case study of Kullu district* (Unpublished Ph.D. Thesis, Himachal Pradesh University, India).
Sood, J. (2016). Homestays in Himachal State, India: A SWOT analysis. *Journal of Tourism*, *17*(2), 69.
Sood, J., Chhabra, D., & Andereck, K. (2017a). Sustainable promotion of homestay tourism in the remote Himalayas of India. Travel and Tourism Research Conference, June, Waterloo, Canada.
Sood, J., Lynch, P., & Anastasiadou, C. (2017b). Community non-participation in homestays in Kullu, Himachal Pradesh, India. *Tourism Management*, *60*, 332–347.
Su, M., Wall, G., & Xu, K. (2016). Heritage tourism and livelihood sustainability of a resettled rural community: Mount Sanquingshan World Heritage Site, China. *Journal of Sustainable Tourism*, *24*, 735–757.
Tao, T. C., & Wall, G. (2009). A livelihood approach to sustainability. *Asia Pacific Journal of Tourism Research*, *14*(2), 137–152.
Thapa, B., & Malini, D. (2017). Guest reasons for choosing homestay accommodation: An overview of recent researches. *Asia Pacific Journal of Research*, *1*(LV), 169–175. Retrieved on October 2, 2019, from https://apjor.com/downloads/2709201728.pdf.
Tompkins, E., & Adger, W. N. (2004). Does adaptive management of natural resources enhance resilience to climate change? *Ecology and Society*, *9*(2).
Wunder, S. (2000). Ecotourism and economic incentives – An empirical approach. *Ecological Economics*, *32*(3), 465–479.

8 Negotiated authentication of nation branding

This chapter offers a discourse on authenticity and authentication of heritage in the context of national branding. Toward the end of the chapter, I will return to the challenges associated with the rebranding of national image to revive and boost heritage tourism during the post-covid times.

A national brand strategy has to resonate with the collective activity of a country, region, and city. A national brand performs an important function as it shares the core value and plans of responding in arduous and challenging times of disequilibrium. Consumers are watching how countries respond to the pandemic and the perceptions that are formed during the pandemic can stay even after the crisis is resolved. Negative perceptions can influence travel decisions later.

According to Dinnie (2020), while COVID-19 is omnipresent, the bigger picture is about health systems and the policies that might shape, in an impactful manner, branding strategies, and future plans of Destination Marketing Organizations (DMOs). Small bed and breakfast (B&B) can get overlooked in recovery efforts and ripple effects of COVID-19 might trickle to developing countries that might not show big numbers but might still suffer from risk warnings for future visitors. Time will tell, how deeply COVID-19 impacts the culture of countries across the globe. Nevertheless, interest in othered culture and heritage will likely to remain ignited and heritage tourism will bounce back. Therefore, it is important to examine how these assets have enriched the image of nations and how they will continue their important role in the future. This chapter offers a discourse on authenticity and authentication of heritage in the context of national branding. Nations are proud to showcase and capitalize on their cultural and heritage treasures as they make them unique and competitive. These treasures are becoming an asset as nations across the world are struggling to redefine themselves as tourist destinations and rebuild their brand identities due to the collapse of international travel. Toward the end of the chapter, I will return to the paradox of rebranding of the national image to revive and boost post-covid heritage tourism.

The extent to which the authenticity of local/regional culture/heritage of a country is modified to showcase a preferred image of a nation forms an important area of study. Culture and heritage are important attributes that tangibilize images of a nation and make it unique and distinct. Literature to date, centered on the role of authenticity and its authentication in national branding, remains scarce although numerous studies have examined the notion of nation branding in the context of generic tourism. Having said that, strong inferences can still be drawn, from documented literature, to carry forward the dialogue of negotiated authenticity and its authentication in the context of nation branding. This chapter first discusses the significance of nation branding and then highlights the importance of authentic cultural/heritage attributes in branding a nation. It then situates nation branding in the context of negotiated authenticity and shares insights on the manner in which heritage attributes are being authenticated. Numerous examples are offered of how culture/heritage attributes are selected, manipulated, and showcased for tourism purpose, thereby dubbing the case for negotiated authenticity. The chapter closes with a brief commentary on the ways nations can rebrand themselves for the post-covid times. Key points communicated in this chapter are:

- Significance of nation branding;
- Issues associated with societal amnesia and dissonance in defining national heritage and their impact on authenticity and the authentication process; and
- Negotiated authenticity and its authentication process in national branding.

A nation is not a product in the conventional sense. It encompasses a wide variety of factors and associations. Anholt (2002) presents a successful national brand hexagon model based on six dimensions: exports, authentic culture and heritage, people, investment and immigration, tourism, and governance. Along similar lines, Fan presents a list of attributes (2006, p. 7):

- Place – geography, tourist attractions
- Natural resources, local products
- People – race, ethnic groups
- History
- Culture
- Language
- Political and economic systems
- Social institutions
- Infrastructure
- Famous persons (the face)
- Picture or image

Evidently, authentic representations of nations are subject to ongoing scrutiny by social scientists because language, historic, ethnic people, natural and local resources, and culture are among the key defining attributes of a nation. Undoubtedly, positive and negative stereotypes exist in different countries and these can impede effective branding efforts, international image (for instance – by projecting a country as dull, primitive, and unrealistic) and financial investment prospects (Avraham 2009). According to Domeisen (2003), a positive national image is key to a flourishing export industry and lucrative investment opportunities. Tourism growth at the national level can be strategically facilitated by developing a differentiating brand. A well-grounded marketing identity should be informed by a sound comprehension of how a country is perceived abroad, while at the same time ensuring that the nationalized attributes are grounded in authentic reality. Competitive positioning on the global map requires strategic dialogue and negotiation of objectively authentic expressions drawn from local and regional cultures. This point is exemplified in the following:

> France, as a nation had had five republics, two empires and at least four kingdoms. France has been royalist, republican and imperial. It has been egalitarian and absolutist in turns and, occasionally, even at the same time, always with the same vigour, sense of destiny and intellectual conviction which distinguishes the French political and cultural scenes. Each time the reality has been modulated, the symbolism has changed with it. And each time, France has presented itself to continue to demonstrate many traditional characteristics. Nevertheless, the changes are not superficial or cosmetic or meaningless, they are real and profound. The reason why nations explicitly and implicitly continue, to rebrand themselves, is because their reality changes and they need to project this real change symbolically to all the audiences, internal and external, with whom they relate. (Olins 2002, pp. 243, 245)

The exit of the British rule led to the creation of new countries like Pakistan and Bangladesh. Bangladesh was first part of India and referred as 'East Bengal', then it became part of Pakistan and was called 'East Pakistan' and now it is called Bangladesh. All of these new countries broke away from their colonial past and, in the process, they unveiled, found, or fabricated a precolonial heritage (Olins 2002). Some other examples of how nations are branding themselves strategically from a global perspective can be evidenced in the national branding images projected by Scotland, Myanmar, and Russia. Scotland has moved away from branding itself as a heritage destination; it now showcases itself as innovative with something to offer to everyone. Its branding message aims at forging an emotional bond by describing itself as friendly, passionate, magical, unique, and welcoming. It promises a mix of natural and cultural experiences:

> Only in Scotland will you find a friendly, passionate and innovative country filled with unique experiences. From the beat of a music festival to the adrenaline of an outdoor adventure, or the breathtaking

> drive through our landscapes to the magic of exploring new attractions, Scotland is waiting to welcome you now. (Visit Scotland 2019)

Myanmar has also rebranded itself and moved away from its image as a cultural destination. It describes itself as exceptional, spiritual, magical, welcoming, and alluring:

> After five years, Myanmar is replacing its tourism branding – Let the Journey Begin – with 'Be Enchanted'. The new brand portrays Myanmar as a friendly, charming, mystical and as-yet-undiscovered destination.
>
> 'Be Enchanted' the tagline is as much a promise as it is an invitation. It's a realization. It's a memory. It's a moment. The word 'enchanted' holds within it the true heart of Myanmar. (Prnewswire 2019)

The new brand: "Visit Myanmar and be enchanted" was developed to prompt curiosity and convey its uniqueness on the global map. The core idea of the rebranding message was taken from a visitor survey at one of its airports. The report suggested that the tagline "Be Enchanted" reinforced pleasurable, warm, welcoming, and satisfactory experiences the tourists have with local hosts. The images evoked in the mind of most tourists were: unique and mystical/mysterious, and experiencing the unknown and the 'other' with one's own eyes. Moving to another country, Russia depicts vastness and a strategic blend of its past heritage and its proud preservation, innovation, and rich resources in its branding image. Brand Russia strives to offer a piece of an objectively authentic past while connecting with contemporary times and competing globally (Understand Russia 2020).

Nations are continually shaping their images to remain abreast and competitive in the global market. Some nations identify appealing cultural and heritage expressions and uplift them on a national pedestal; in this manner, they negotiate their authenticities to generate a preferred national identity. Some nations, on the other hand, have disregarded their cultural and historic resources and have been more focused on portraying themselves as economic and political powers. For instance, Lieven (2000) talks about nascent German nationalism and states that this doctrine overshadows the romanticist image of Germany, in that it compromises ethnicity and language that epitomize the core identities of local communities. Spain rebranded itself to portray as authenticity instead of holding on to stereotype images/perceptions that depicted it as an economically backward and isolated nation under a dictator rule (Olins 2002). In rebranding its image, it focused on showcasing its authentic cultural resources and local heritage. Other examples of reinvented nations include Ireland and the former communist countries (Eastern Europe).

Numerous issues with regard to national branding are shared in recent literature. It has also been argued that nationalism can be prosaic and unnoticed and subconscious practices can generate authentic identities, thereby supporting the notion of spontaneous nation building (Polese & Horak 2015). Many times, there is little overlap between local, regional, and national interests in national branding although a shared sense of purpose or objective can facilitate meaningful cooperation. Ordinary citizens can engage in building national images. Therefore, all versions of national identities are not constructivist and do not necessarily get compromised by political agenda. Pawlusz and Polese (2017) use Estonian experience, as an example of building a nation for the purpose of tourism, to demonstrate that a national identity can be negotiated and formed in an authentic manner:

> by analyzing the way official tourism brochures, Estonian tourism narratives in general, and the way of 'selling' the country to tourists have developed, we will show a desire to 'convince the others to convince ourselves.' Attention is on the role of culture/heritage in nation-building efforts of post-Soviet countries. Tourist rhetoric often claims the existence of a collective identity and well-defined traditions organized around geographical sites and practices. It also conveys the sense that the present must preserve its links to a(n) (often romanticized) past that seems to be vanishing in the globalized and modernized world. (p. 874)

Table 8.1 illustrates how heritage tourism can inspire the manner in which a nation can brand itself. Pillars of Estonian nation-branding are founded on positive transformation by offering a welcoming and positively surprising image as 'the best kept secret of Scandinavia'. The brand image boasts of natural ties with Scandinavia and revolves around four cornerstones of Estonia to portray its authenticness: the Nordic influence and rootedness; the eastern influence and progress. Pawlusz and Polese (2017) present a marketing concept model for Estonia tourism and center it around two dimensions: heart and soul. The target markets of focus are immigrants, businesses, foreign, and domestic tourists. According to the authors:

- Estonia's authentic Nordicness and eastern influence make an attractive package for immigrants, whereas, the business actors can be wooed by Estonia's progressiveness and Nordic quality. Easternness is transmitted as an economic asset (low prices), whereas "Nordicness" is presented as a cultural and social attribute;
- The 'four cornerstones of identity' are portrayed as contrasts that together contribute to Estonia's uniqueness and create the feeling of surprise. Estonia's main slogan, "positively surprising", is aimed at building a differentiated image and removing the stereotype image of an underdeveloped post-communist country;
- The ideas of surprise, tension, or contrasts are rooted in the perception of the location of Estonia, Poland, or Romania at the crossroads of 'East' and 'West'. This perceived tension is turned into an added value

Table 8.1 Nation branding through heritage tourism

With regard to Estonia, strategic initiatives are centered on launching the legitimacy and 'authenticity' of identities belonging to the pre-Soviet era while also connecting them in a metaphoric manner to the post-Soviet times. The tourism expression, therefore, orbits around the symbolic ethnic past and the purpose is to inculcate a sense of historical and cultural intransience for the nation. This is often in congruence with the efforts to discard, ignore, or brand the Soviet past as undesirable.
Creation, treatment, and views of Estonian cultural heritage have been, for the most part, shaped by two components. First, the nation building process is engrained in Herderian philosophy of cultural nationalism (Smidchens 2014), which was developed in the late nineteenth century and established in the period of independence between the two wars. Despite the denationalization goal of the Soviet cultural policies, an ethnic segregation was promoted between the Soviet society and the features mirroring the Estonian ethnos, thereby supporting cultural nationalism (Seljamaa 2012). These two factors influenced the post-independence policies in a predominant manner and led to the creation of an official hybrid national storyline that stressed on detachment from the Soviet past and rediscovery of the Estonianness by the country and its people. This laid the foundations for fueling anti-Russian expressions (Vetik 1993) and also became a starting point for initiatives focused on building a 'new' Estonian identity using tourism and branding campaigns as a platform.
Tourist brochures in Estonia, while emphasizing selected features of the national culture and celebrating the richness of the country, communicate an idea to augment stories of national identity. The invention of Estonian 'traditions' and the official descriptions of Estonian identity are reconfigured by collaborations between state and non-state agencies and also by day-to-day practices of the Estonians (Seliverstova 2017).

Source: Pawlusz and Polese (2017)

that makes those countries different yet familiar to both the Western and domestic audience. Estonia's Nordic heart – elegant, clean, and simple – is juxtaposed with its Eastern soul – hospitable, exotic, and spontaneous; and

- Rootedness and the Nordic influence are further pictured as primary juxtaposition of the modern and the sophisticated West with the more traditional and spontaneous East with a "Russian" or "Slavic" hospitable soul (Pawlusz & Polese 2017, p. 878).

From the aforementioned descriptions, it can be seen that Estonia has strategically rebranded itself to portray a hybrid image. That is, a strategically negotiated mix of authentically rooted cultural experiences and economical services are marketed to demonstrative a unique setting offering economically accessible services and experiences.

Pawlusz and Polese further write that heritage and the past circumscribe the Estonian brand. The branding campaign is dedicated on conveying, to domestic and international visitors, a sense of rootedness "in ancient culture and traditions, the unique language, and vibrant folk heritage. This echoes post-Soviet nation-building, which exposed national heritage in an attempt to reestablish a sense of identity and national pride (Picard 1997). In contrast to the young and still fragile state, culture serves as a validation and anchor of the nation's ancientness, timelessness, and therefore, Europeanness before the Soviet invasion. The ancientness of the country and its rural charm are further juxtaposed with its progressiveness, fast technological development, and widespread access to the Internet" (2017, pp. 874–879).

Dzenovska (2005) writes that a preferred narrative, with a collective identity that stimulates self-esteem, forms the foundations of an authentic national image of Estonia. For instance, the design of Tallinn Airport revolves around sense of local 'home' and homeland. Commercial areas such as cafes, shops, and restaurants are designed to make customers feel a sense of home (Dzenovska 2005). Objects such as armchairs are reminiscent of home furnishings and added soft draperies convey a sense of comfort and homeliness; Table legs are made of raw birch branches from local trees; furniture for passengers is decorated with folk designs, from different regions of Estonia. The environment and its ambience is made appealing to both tourist gaze (by showing something novel and unique) and local gaze (by offering a local sense of relaxation is generated with the use of local resources and language).

National heritage is highly negotiated because its authenticity is subject to preferred political ideologies. For instance, Pretes postulates that selected heritage sites that showcase a nation's past are used by ruling governments to construct a national identity and "in the US, both government and private resources helped shape a common 'American' identity from diverse immigrant backgrounds and heritage sites have played an important role in this identity formation" (2003, p. 126). He contends that nations are imagined communities and their authentic showcasing is negotiated and orchestrated to transmit preferred messages. Table 8.2 presents an illustration of his argument. Furthermore, as explained by Timothy and Boyd (2003), it is common for officiating governments to use heritage to structure public views, to foster nationalist sentiments, and to build images that portray their political ideologies.

According to Pretes, "in the United States, cultural production – in such forms as celebrations, monuments, and sights – is essential in creating a national identity" (2003, p. 128). The author provides three examples of cultural attractions in South Dakota and illustrates how they instill a preferred feeling of authentic national identity: Mount Rushmore National Memorial; Wall Drug Store; and Rapid City Dinosaur Park: Science and Nature. At Mount Rushmore, four of the nation's greatest presidents are sculpted and memorialized: George Washington, Thomas Jefferson, Abraham Lincoln, and Teddy Roosevelt. The monument instills a feeling of 'National Identity'

Table 8.2 Nations as imagined communities

The construction of an imagined national community is reliant on national institutions to communicate a sense of shared identity. Museums, by legitimizing history of a nation and inculcating a sense of shared heritage, showcase the core characteristics of nationhood and substantiate its historical existence. Furthermore, akin to the museums, other role players are archeological sites and their reconstruction marks the foundation apologue of the nation through sharing of deep history and connection with the land. The archeological practice involves five phases: "first, the writing of reports; second, the production of illustrated books; third, logoization or the display of archeological image on postcards, postage stamps, and so forth; and fourth, the market or the use of the image as a marketing device" (Andersen 1991, p. 182). The fifth phase is linked with tourism, when tourists perform the role of pilgrims as they visit the 'archeological' sites.
New archeological sites were constructed because the Native American sites did not convey the preferred message to the American audience. Sites such as Mount Rushmore were given the same status as the Pyramids in Egypt or the Parthenon in Greece, as symbolic markers of national identity and heritage. These monuments are enshrined on the national currency, postage stamps, and other legitimized products besides being commemorated in commercial advertising, entertainment, and the media. They depict a shared value and sense of identity thereby popularizing an authoritative national expression of inclusion.

Source: Pretes (2003)

and communicates the national contribution of each president by engraving their work on the mountain:

> As an instance, Washington led the country to its birth, Jefferson doubled it in size, Lincoln saved it, and Roosevelt showed us how the common man can stand up against corporations. The nation's first president led the charge against an oppressive regime; Jefferson represents endeavor and bravery, as he purchased a chunk of land that would lead to the exploration and civilization of an entire half of a continent, and not a calm one, on the wings of manifest destiny; Lincoln represents equality, of course, as his mental fortitude would see this nation through its bloodiest conflict, an internal one, in which he unified an utterly shattered union; Roosevelt represents a more modern United States as a hub of global corporate power. (Pretes 2003, p. 128)

With regard to the Wall Drug Store, it showcases an entrepreneurial success story and proudly conveys authentic American values through: success of private businesses; free enterprise, boldly surviving the Great Depression, innovative; protestant work ethic; and prosperity through hard work and perseverance (Pretes 2003). Last, dinosaurs are an American animal because they were strong and superior in size. According to Kirby, Smith, and Wilkins (1992), an American patriot shares a strong sense of identity with the dinosaurs. The American culture identifies with dinosaurs and regards it as its national animal. Fossils of dinosaurs signify the archeological finds of the American past.

Boundaries of nations are endorsed by maps and favored/fancied images of shared heritage. In line with this, several countries have introduced new versions of heritage to redefine shared nationalist ideals. This compromises the objective authenticity attributes and has often resulted in heritage dissonance at the local levels. Many times, communities at local and national level disagree with "what heritage to preserve, value, and incorporate into their identity" (Frost 2006, p. 6). Howard holds that while "heritage always benefits someone .... (it also) disadvantages someone else" (2003, p. 4). For instance, the 1988 Bicentenary organizers in Australia had made efforts to exclude indigenous groups. Therefore, the issue is not of multiple heritages and their authenticities at a particular point of time; but it is about which story to tell and whose heritage to showcase. Heritage dissonance (selected heritage for display) and intentional amnesia will continue to be a contentious issue. Intentional amnesia means deliberate forgetting of some objectively authentic aspects of the past and history, which refers to facts that societies prefer to disregard, exclude, or suppress because they are uncomfortable, embarrassing, or by doing so, the party in power can achieve some political or ideological objective often with a racist slant.

Many colonial superpowers and autocratic administrations have granted independence to their former colonies, paving the way for the emergence of several democratic and capitalist nations. Upon achieving freedom, some of these new nations have expressed a desire to erase their colonial past thereby erasing old records to forget memories that are hurtful, oppressive, and embarrassing. Rewriting of history also highlights the meaning of 'absence' of some aspect of history and objectively authentic heritage. The overall impact, of this shaping of memory through the erosion of former histories, raises the question of accurate history or complete absence of history. According to Timothy (2011), it is a partial and selective version of history that denies others their place in history. History is, therefore, a politics of 'absence' that is collectively embraced; that is, it is not only the tangible reminders of the past that help to forge a national identity, but also that which is invisible or absent (Frost 2006). As a case in point, I have illustrated several examples such as Estonia, Yugoslavia, Croatia, and India. Struggles, between ruling agencies, are often focused on "how much of a heritage should be removed and what should replace it" (Timothy & Boyd 2003, p. 263).

Hall (2002) studies the manner in which states branded themselves after the disintegration of Yugoslavia. The political division of former Yugoslav, which once showcased a unified and harmonious image led to each independent state striving to put behind its communist past and craft a new national image and identity. Hall (2002) examines the branding of former states, which are using tourism to help portray a preferred national identity and place them on a revitalized economic path. The author notes that the former Yugoslav, "with its increasingly complex series of two-way linkages with Western Europe, and notably with German-speaking countries, provided the context within which Yugoslavia proceeded to cement a national identity in the West around experiences of inexpensive coastal mass tourism reinforced with the images of such cultural-historic icons as the medieval walled city of Dubrovnik" (Hall 2002, p. 324). Furthermore, he writes that Serbia under the Milosevic regime is a notable example of the manipulation of ethnic identity for callous political wins. While in rural regions, the images of heritage are used to facilitating restructuring and modernization, few national and regional governments have manipulated the heritage industry to fortify a preferred national or ethnic identity. As an example, in the promotion of Serbia:

> landscape was interpreted as both natural and cultural. In terms of heritage, however, the images of cultural 'landscape' which were portrayed revealed an exclusive representation of Serbian/Orthodox tradition. One third of the total population of Serbia was not Serbian at this time, but in the document there was no mention of minority ethnic Hungarian or Albanian Catholic or Muslim heritage" (Hall 2002, pp. 331–332).

The former Yugoslav states present a notable example of national branding initiatives as they were enveloped by political/economic events. A close association between tourism branding and national identity exists with an interplay between economic and political agenda, thereby shaping remnants of objectively authentic culture and history. Hall captures the reimaging efforts of some of these states, especially in the context of their negotiation of history and heritage:

- The escape strategy from the past and its images has partly aimed to move away from the image of mass coastal tourism and to highlight the novelty of its cultural and natural resources;
- Heritage occupies a more ambivalent role. In a region with a diversity of cultures and histories and where the past is often drawn on to justify the present, 'heritage' tourism offers the irony of employing the past as an element of restructuring for the future, particularly for the newly independent states on their present communist roots;
- Heritage is far from being a value-free concept – economic power and politics influence what is preserved and how it is interpreted. Promotion of secular rural and urban heritage as an integral element of cultural history was characteristic of the communist period. Under the communists, however, such heritage was employed to overcome or subdue ethnic rivalries and as a means to inculcate a unified sense of Yugoslav identity and pride; and

- The resurgence of nationalist expression alongside of (re-)creation of new state systems has encouraged some countries to employ the heritage industry as a means of reinforcing national or a particular ethnic identity (Hall 2002, p. 325).

In the context of Croatia, the aftermath of the war brought not only physical dislocation prompted by the redefinition of geographical space, but 'also cognitive dislocation – a disjunction between the concept of home in the past and in the present' (Mountcastle & Danon 2001, p. 107). Feelings about other members of the former Yugoslav Federation have been negatively showcased. Several examples of modulating the positive identity of Serbs can be noted in words and narratives such as: Bred for battle; Serbs are born with a blade beneath their pillows; You can tell a Serb by the way he walks; and feelings of anger and desire to destroy all historic Serb artifacts (Goulding & Domic 2009). Such feelings provoke unfair judgments by the current generation who had nothing to do with the war. Tradition and history are matters of how powerful institutions function to select values from the past and mobilize them in contemporary practices (Wright 1985). Essentially, cultural reproduction can influence collective memory and result in a particular sense of national and cultural identity (Morley & Robbins 1989). Along similar lines, Edensor (1997) points out that attempts, to fix national memory and identity and mapping a territory with preferred history, are integral to the ideological rhetoric of nationalism.

As illustrated, Croatia has mapped its identity and tried to distance itself from its Serbian past. In order to "create" a new history and sanitize its past, Croatians have compromised parts of its objectively authentic heritage and eliminated items/sites that are rich in Serbian heritage. Furthermore, the Croat government is focused on showcasing the past, present, and future that solely reflects the Croat culture which is negotiated to transmit preferred values. Language has also been manipulated and the language change, as an outcome of political agenda, has not been welcomed by the local communities (Goulding & Domic 2009). Goulding and Domic comment that "in Croatia, the past, the present and the future are inextricably interconnected as the country works towards reinventing itself as 'independent', autonomous, and above all Croat" (Goulding & Domic 2009, p. 93). New Croatia accomplished this by physically destroying all 'partisan remnants' from its time of conflict, and utilized the media to promote a politically negotiated image of Croatian heritage.

As illustrated earlier, these messages strategically eliminate mention of Serbian involvement during the clash and focus only narratives associated with Croatian independence and bravery. Croatian festivals and displays portray war-time reenactments of Croat victories, advocating the image of Croatia as a strong, independent nation-state not a victim of violence. The Croatian government has even taken steps to eliminate or modify certain elements of its language to distance itself from its pre-war situation. Croatia has shaped its history using societal amnesia (Goulding & Domic 2009).

Slovenia, on the other hand, has tried to negotiate its cultural identity by stressing on: "a fashionable/politically correct ecologically friendly ethos; its Central European character, Habsburg Heritage, Alpine associations and contiguity with Austria and Italy; and disengagement from any 'Balkan' connotations" (Hall 2002, p. 329).

Yet another striking case in point is Brand India. Mehta-Karia (2011) presents a notable discourse on the story of nation branding of a postcolonial country who changed her reputation of being poor and positioned herself in the epicenter of the global economic gaze. Mehta-Karia's (2011) thesis questions the narrative used to make India a favorite among the global corporate entrepreneurs at the cost of compromising its rich cultural and rural identity. She contends that "India imagined through the Brand is an India that valorizes particular discourses, exalts particular narratives and fetishes particular subjectivities. This process of exalting subjectivities promoted by the brand entails the marginalizing of those that can't and won't fit in" (2011, p. 5). In retaliation, a leading national newspaper (Times of India) launched an 'India Poised' multimedia crusade to raise awareness of the under-showcased 'othered' India. It was a twelve-month long endeavor orchestrated by a leading actor of Indian cinema (Bollywood), Amitabh Bachchan. The actor delivered a poetical narrative of the 'India Poised' anthem, written by Gulzar, a famous Bollywood lyricist (Mehta-Karia 2011, p. 6). The lyrics of the poem – India vs India offer remarkable insights on othered versus modern India (Gulzar 2007, p. 1):

> There are two Indians in this country. One India is straining at the leash, eager to spring forth and live up to all the adjectives that the world has recently been showering upon us. The other India is the leash.
>
> One India says, 'Give me a chance and I'll prove myself'. The Other India says, 'prove yourself first and then maybe you'll have a chance.
>
> One India lives in the optimism of our hearts. The Other India lurks in the skepticism of our minds.
>
> One India wants. The Other India hopes. One India leads. The Other India follows.
>
> These conversations are on the rise. With each passing day, more and more people from the other India are coming over to this side. And quietly, while the world is not looking....
>
> A pulsating, dynamic new India is emerging. An India! Whose faith in success is far greater than its fear in failure. An India that no longer boycotts foreign-made goods, but buys out the companies that make them instead.
>
> History, they say is a bad motorist. It rarely ever signals its intentions when its taking a turn.

> This is that 'rarely-ever' moment. History is turning a page. For over half a century, our nation has sprung, stumbled, run, fallen, rolled over, got up and dusted herself, and cantered, sometimes lurched on.
>
> But now, in our sixtieth year as a free nation, the ride has brought us to the edge of time's great precipice. And one India, a tiny little voice at the back of the head, is looking down at the bottom of the ravine and hesitating. The Other is looking up at the sky and saying, 'Its time to fly'.

This chorale of 'New India questions the 'animated', 'vibrant', and the 'innovative' identity versus the 'Othered' (objectively authentic India) India that is less privileged, neglected, and absent from the global discourse and showcasing. In the global world, the 'New' India is desired and rejoiced whereas the 'Other' India is an undesired baggage (Mehta-Karia 2011). In this context, Mehta-Karia makes a logical argument:

> While analyzing the trajectory of the nation-state as imagined through the discourse of branding, it thus becomes imperative to ask, how is it that the complex and contesting subjectivities of a billion plus citizens of the Indian nation came to be crystallized by a group of men over a hotel breakfast discussion? What does it mean when democracy is collapsed with free market in order to make the nation more attractive to the global marketplace? What kind of subjects are invited to participate in this discourse? And alternatively, what subjects are silenced, what subjectivities obscured and erased to sustain and legitimize this course? (2011, pp. 9–10).

At this juncture, I would like to draw attention to the India-Brand Equity Foundation (IBEF). It is a public-private sector syndicate that has played a key role in shaping the brand vision of India. Its identification symbol is a cageless soaring bird, with a green wing and a saffron wing matching the colors of the Indian national flag thereby implying a free nation with the ability to soar upward into the boundary-less sky; evidently, it is these type of symbols that brand a nation rather than the national flag that might communicate historical baggage and political connotations (Mehta-Karia 2011). This shows the core purpose of national branding, for many countries, is the promotion of global trade and corporate entrepreneurship to competitively position themselves in an already crowded global marketplace (Mehta-Karia 2011).

To what extent these kind of branding goals influence negotiated authenticity and its authentication of unique cultural and heritage assets forms an interesting area of study as an overtly economic agenda is antithetical to what objective authenticity or sustainable negotiated authenticity perspectives/advocates espouse. UNESCO, for instance, confers world heritage status to unique and fragile sites. But it is not aimed as a marketing initiative and commercialization tool although countries have strategically submitted bids for their fragile and novel heritage sites to boost national branding and commerce. This interplay between the local and universal is evident in heritage tourism. In rare instances, the local custodians of a heritage site are powerful enough to counter the 'economification' of national branding; otherwise, the political agencies exploit it for commerce and to promote mass tourism.

Closing comments

Volcic (2008) points out that nation branding strategies commodify nations' histories and bestow an 'othered' image of its citizens that appeals to the West. Klein (2002) has argued that nation branding contradicts the core substance of democracy and democratic policies. Aligned with this view, it is postulated that the nation branding exercise shrinks the public space by showcasing the country from a purely economic perspective (Jansen 2008; Mehta-Karia 2011). From a moral perspective, some scholars question nation branding as the authority moves from the political sphere to the marketing consultants and business entrepreneurs (Araonczyk 2008; Mehta-Karia 2011). According to Mehta-Karia, "nation branding is a practice that facilitates the creation of conditions that introduce, promote and naturalize 'market solutions' and 'enterprise culture' for all aspects of society – be it individual, personal or national, economic or social" (2011, p. 4). Negotiated authenticity of heritage for tourism purpose and its authentication path, in such cases, is driven by political and economic agenda.

Pawlusz and Polese argue that the contemporary era seeks "nostalgia for authenticity, and tourism is a means to access the 'real', 'authentic' culture of the self and the other"; while at the same time, authors illustrate that "the sense of national belonging and national markers are not constants to be exploited by the tourism industry, but rather they are constantly remade, retold, and popularized in the course of tourism development" (2017, p. 884). During the pre-covid era, the notion of nation branding dictated by the global marketplace was raising apprehension, with regard to the problematic manner it was influencing the heritage institutions across the globe. Also, troublesome was the manner in which they authenticated authenticity to showcase preferred heritage of a nation. Evidently, leaning has been toward a negotiated position authenticated but with a favored selection of markers that appeal to the global markets and the preferred west. For instance, negotiated authenticity process to appeal to the western audience is clearly evident in the authenticating markers of nation branding and images used by Eastern Europe and India. Negotiated commodification of a nation's brand forms a notable field of inquiry to assist in gathering insights on the social-constructivist and political nature of authentication.

The authentication process of nation branding, in the context of heritage tourism, needs to identify different role players and unmask their social constructivist/political settings to understand the manner in which they are interrelated so that they can effectively coordinate with each other. In doing so, a dedicated effort can be made to expose the burden of illusionary as well as capitalist fallacies imposed by the global corporate culture or preferred political ideologies on national branding initiatives and expressions. Also, dissonant heritage narratives need to be resolved in a manner that contradictions and controversies can be showcased side by side, for instance, by building cultural memory routes. Such multi-perspective representations "have attractiveness and give the possibility of participation (at least intellectual curiosity participation) in imaginary creation (civic engagement), for both domestic inhabitants, regional and foreign tourists" (Sesic & Mijatovic 2014, p. 17).

The overarching aim should be to authenticate pluralism harmoniously, dissolve the conflict between contradictory representations, and make heritage resilient. In this manner, dissonance can be mitigated by giving voice to both silent and vocal communities so that they are able to showcase their views of an 'othered' nation outside of the branding narratives. For instance, in the context of India, Mehta-Karia shows that the marginalized communities "imagine an India where the 'Old' is not discarded and the 'New' is not imposed" (2011, p. 105).

The pandemic has changed the core tenets of national branding. Future branding strategies will need to project the image of a reinvented or transformed nation. According to Torres (2020), a national branding strategy should consider the numerous perceptions that make up a countries' image in the minds of people, that is: the way in which governments manage a country, the identity, and culture of the country as well as its history and territory. Over time the combination of these perceptions builds an image related to the nation brand that results in an emotion. This emotion impacts the citizen's willingness to live, study, work, visit, invest, and buy in that country. National brands are only affected by crises when they are intense and prolonged and have an impact on the identity and culture of a country, and the current pandemic has already consumed three-fourths of the 2020 year. This implies, that every nation will have to reinvent its image to make a comeback on the international tourism map.

Heritage tourism is one platform that can showcase silent or forgotten voices and stimulate sustainable development to promote quality of life and well-being of host communities. Its 'authenticness', unique to its people, topography, and culture, can help a nation to reinvent its image during times of unpredictable crises. As explained by Sesic and Mijatovic (2014), heritage tourism holds enormous potential to negotiate alternative cultural routes to resonate with changing times. It can also reauthenticate, make a national brand culturally resilient, and globally position it as a repository of unique history and enriching cultural diversity. The crucial role of museums, as significant authenticating agents of national heritage representations, cannot be denied in this regard. The next chapter deliberates on their position.

References

Andersen, B. (1991). *Imagined communities: Reflections on the origin and spread of nationalism* (revised ed.). London: Verso.

Anholt, S. (2002). Nation branding: A continuing theme. *Journal of Brand Management, 10*(1), 59–60.

Araonczyk, M. (2008). Living the brand: nationality, globality, and the identity strategies of nation branding consultants. *International Journal of Communications, 2*, 41–65.

Avraham, E. (2009). Marketing and managing nation branding during prolonged crisis: The case of Israel. *Place Branding and Public Diplomacy, 5*(3), 202–212.

Dinnie, K. (2020). Crisis Management and COVID-19. Retrieved from https://www.citynationplace.com/crisis-management-covid-19-pandemic-impact-on-nation-brands.

Domeisen, N. (2003, January). Is there a case for national branding? In *International Trade Forum* (No. 1, p. 14). International Trade Centre.

Dzenovska, D. (2005). Remaking the nation of Latvia: Anthropological perspectives on nation branding. *Place Branding, 1*(2), 173–186.

Edensor, T. (1997). National identity and the politics of memory: Remembering Bruce and Wallace in symbolic space. *Environment and Planning, 29*(2), 175–194.

Fan, Y. (2006). Branding the nation: What is being branded? *Journal of Vacation Marketing, 12*(1), 5–13.

Frost, W. (2006). Braveheart-ed Ned Kelly: historic films, heritage tourism and destination image. *Tourism management, 27*(2), 247–254.

Goulding, C., & Domic, D. (2009). Heritage: Identity and ideological manipulation: The case of Croatia. *Annals of Tourism Research, 36*(1), 85–102.

Gulzar (2007). India vs. India. Transcribed September 9, 2009, from http://www.indiapoised.com/video2.htm in Mehta-Karia (2011)C.

Hall, D. (2002). Brand development, tourism and national identity: The re-imaging of former Yugoslavia. *Journal of Brand Management, 9*(4), 323–334.

Jansen, S. (2008). Designer nations: Neo-liberal nation branding. *Social Identities, 14*(1), 121–142.

Kirby, D., Smith, K., & Wilkins, M. (1992). *The new roadside America: The modern traveler's guide to the wild and wonderful world of America's tourist attractions*. New York, NY: Fireside Books.

Klein, N. (2002). The Spectacular Failure of Brand USA. Retrieved from http://www.naomiklein.org/articles/2002/03/spectacular-failure-brand-usa.

Lieven, D. (2000). *Empire*. London: John Murray.

Mehta-Karia, S. (2011). *Imagining India: The Nation as a Brand.* Unpublished Thesis. University of Toronto.

Morley, D., & Robbins, K. (1989). Spaced of identity: Communication technologies and the reconfiguration of Europe. *Screen, 30*(Autumn), 32–59.

Mountcastle, A., & Danon, D. (2001). Coming" Home": Identity and Place in Post War Croatia. Narodna umjetnost, 38(1), 105–19.

Olins, W. (2002). Opinion piece-branding the national – the historical context. *Journal of Brand Management, 9*(4/5), 241–248.

Pawłusz, E., & Polese, A. (2017). "Scandinavia's best-kept secret." Tourism promotion, nation-branding, and identity construction in Estonia (with a free guided tour of Tallinn Airport). *Nationalities Papers, 45*(5), 873–892.

Picard, M. (1997). *Tourism, ethnicity, and the state in Asian and Pacific Societies*. Honolulu: University of Hawaii Press.
Polese, A., & Horak, S. (2015). A tale of two presidents: Personality cult and symbolic nation-building in Turkmenistan. *Nationalities Papers*, *43*(3), 457–478.
Pretes, M. (2003). Tourism and nationalism. *Annals of Tourism Research*, *30*(1), 125–142.
Prnewswire (2019). Myanmar unveils its new tourism branding. Retrieved on October 15, 2019, from https://www.prnewswire.com/news-releases/myanmar-unveils-new-tourism-branding-300713508.html.
Seliverstova, O. (2017). 'Consuming' national identity in Western Ukraine. *Nationalities Papers*, *41*(1), 61–79.
Seljamaa, E. (2012). *A Home for 121 Nationalities or Less: Nationalism, Ethnicity, and Integration in Post-Soviet Estonia*. PhD Dissertation: The Ohio State University.
Sesic, M., & Mijatovic, L. (2014). Balkan dissonant heritage narratives (and their attractiveness) for tourism. *American Journal of Tourism Management*, *3*(1B), 10–19.
Smidchens, G. (2014). *The power of song. Nonviolent national culture in the Baltic singing revolution*. Seattle, WA: University of Washington Press.
Timothy, D. (2011). *Cultural and heritage tourism*. Bristol: Channel View Publications.
Timothy, D., & Boyd, S. (2003). *Cultural and heritage tourism*. London: Channel View Publications.
Torres (2020). Retrieved on July, 2020, from https://www.bloom-consulting.com/journal/what-nation-brands-must-do-to-tackle-covid-19/.
Understand Russia (2020). Retrieved on July, 2020, from https://understandrussia.com/branding-russia/.
Vetik, R. (1993). Ethnic conflict and accommodation in post-communist Estonia. *Journal of Peace Research*, *30*(3), 271–280.
Visit Scotland (2019). Visit Scotland. Retrieved on October 15, 2019, from https://www.visitscotland.com/.
Volcic, Z. (2008). Former Yugoslavia on the world wide web: Commercialization and branding of nation-states. *The International Gazette*, *70*(5), 395–413.
Wright, P. (1985). *On living in an old country: The national past in contemporary Britain*. London: Verso.

9 Negotiated authentication of museums

This chapter examines the contemporary museology in the context of negotiated authenticity, that is, how museums authenticate a tradeoff between their conventional focus on objective authenticity and demand for existentialist/constructivist offerings. Furthermore, a discourse is offered on how this negotiated authenticity is authenticated using the digital backdrop. The chapter closes with an overview of the manner in which museums are operating in the pandemic times.

The museums continue to negotiate the authenticities of their resources to remain in context with their patrons, the hosting communities, and the governing agencies. Undoubtedly, museums are important repositories of past histories. They are revered for their role as agents of conservation and cultural authority. However, in the past few decades, the role of museums has expanded from serving as ivory towers of past heritage to catalysts of community engagement. They are embracing social-constructivist approaches to transform themselves into socially and economically viable cultural institutions. In the contemporary era, the museums are expected to connect with the identities and interpretations of their audience and tap into community intelligence by facilitating "social networking and demand-driven intellectual engagement" with their cultural displays (Russo, Watkins, Kelly, & Chan 2008, p. 22). Because of the changing trends and budget cuts, they have adapted their policies and objectively authentic dispositions to embrace new ideologies and markets. This stance is evident in the mission statements of numerous museums across the globe:

- "The mission of the **Heard Museum** is to be the world's preeminent museum for the presentation, interpretation and advancement of American Indian art, emphasizing its intersection with broader artistic and cultural themes" (Heard Museum n.d.)
- "**The Metropolitan Museum of Art** collects, studies, conserves, and presents significant works of art across all times and cultures in order to connect people to creativity, knowledge, and ideas" (Metropolitan Museum of Art 2019)

- "Our mission is to preserve, interpret and make accessible for all, the past and present of **Scotland**, other nations and cultures, and the natural world" (National Museum of Scotland 2019)
- "**Te Papa** is a forum for the nation to present, explore, and preserve the heritage of its cultures and knowledge of the natural environment in order to: ... enrich the present and meet the challenges of the future" (Te Pa Museum 2019)
- **The Hermitage Amsterdam** aims to use art and history to inspire, enrich, and above all offer opportunity for reflection. We take inspiration from the historical ties between Amsterdam and St Petersburg and between the House of Orange-Nassau and the Romanovs. We believe that innovation and education are essential. As an enterprise, we take pride in the fact that we finance our activities from our own resources, with support from the business community and private individuals (Hermitage Museum 2019).

Evident in the mission statements is emphasis on connecting and engaging with the contemporary audience and enriching the present. Numerous museums across the globe are striving to be innovative and acknowledge the significant role of their tourism stakeholders to enable a complete satisfactory visitor experience. In terms of innovation, technology has emerged as the most significant factor that continues to shape the contemporary museology. Hung, Chen, Hung, and Ho write that "the proliferation of information communication technology (ICT) is transforming all aspects of museum operations while enhancing the traditional functions" (2013, p. 231). Undoubtedly, ICT has played a key role in redefining the new museum ethos of showcasing culture in a manner that its value matches cash. This chapter examines the contemporary museology in the context of negotiated authenticity, that is, how museums authenticate a tradeoff between their conventional focus on objective authenticity and demand for existentialist/constructivist offerings. Furthermore, a discourse is offered on how this negotiated authenticity is authenticated using the digital backdrop. The chapter closes with an overview of the manner in which museums are operating in the pandemic times.

Key highlights of this chapter include:

- Museums and their new museology
- Digitalization of museums and negotiated authenticity
- Authentication of negotiated authenticity, especially in digital settings

The new museology

Museums showcase both intangible and tangible heritage. Capturing many notions of museums, Burcaw defines a museum "as a nonprofit institution that collects, preserves and displays objects for educational or aesthetic purposes" (1975, p. 9f). Museums not only store physical objects but also communicate knowledge (Schweibenz 1998). Because of dwindling public funds, they are expected to stretch their visitor portfolio by stretching beyond their role as agents of conservation; instead they are expected to offer an authentic experience that is mediated through objects and curatorial interpretation with an intention to inform, please, and prompt interest/engagement (Britton 1991; Chhabra 2007, 2010; Harrison 2005; Prentice 2001; Stephen 2001). Supporting this line of views, Pearce argues that "museums have always been, and are still, deeply implicated in the maintenance of the capitalist market system" and they play an important role "in authenticating and projecting a clear hierarchy of value in which cultural values match cash" (1992, pp. 235, 237).

Ongoing democratization have shifted their focus from "serving as cabinets of curiosity in the 17th century to emphasizing education in the 19th century to public empowerment in the 20th century" (Graburn n.d.). Clearly, museums have evolved and pursue multiple ideologies to keep their resources meaningful in the present times; they have not remained homogeneous in the face of changing sociopolitical and economic environments. MacDonald and Alsford attribute this shift to the argument that "all museums are products of their particular cultural and historical experiences" (1995, p. 24). Russo, Watkins, Kelly, and Chan capture this dynamic museum nomenclature in their narrative:

> Museums are increasingly open to cultural diversity, local knowledge, and popular memory. Social constructivist approaches to communication have helped museums to connect with the memories, identities, and understandings that visitors bring with them (Hein 1998; Watkins & Mortimore 1999). The same approaches have enabled the deconstruction of grand narratives and have affirmed the role of audiences in social learning. These debates have tapped a form of community intelligence and have created a path from modernist certainty and institutional centrality, to social networking and demand-driven intellectual engagement with culture. In turn, this has changed the ways that museums respond to the challenge of providing authentic and authoritative information within an increasingly participatory online environment. Museums are now sites in which knowledge, memory, and history are examined, rather than places where cultural authority is asserted (Kelly, Cook, & Gordon 2006; Witcomb 1999). Museums and visitors collaborate in the 'making of meaning' whether visitors are local residents who lived through a particular time, or school students working on problem-based research projects (Hooper-Greenhill 2000; Silverman 1995). Existing studies suggest that museums enable cultural participants to explore images of themselves, their histories and communities (Falk 2006). This shift within the museum has resulted in initial experimentation with social media and participatory cultural communication. (2008, p. 22)

Jones (2010) says that a core message communicated by recent literature on museums is that authenticity is not embedded in an object. Rather, it is a culturally constructed quality that differs based on context and the audience. Belk (2001) also contends that the magnetic spell are the personal meaning and bestowed values. Hede, Garma, Josiassan, and Thyne write that "visitor's subjective interpretations of their own authenticity, as museum visitors, are distinct from their subjective interpretations of the authenticity of the materials in the museum" (2014, p. 1406). Because consumers often employ self-oriented behaviors or consumption associated with their extended self, their evaluations may differ (Beverland, Lindgreen, & Vink 2008). Hede, Garma, Josiassen, and Thyne (2014) suggest that "self-authentication may operate at two levels: initially on the level of the consumption of the market offering, when consumers situate themselves in a specific consumption context, and then at the gestalt level, when they evaluate how a specific consumption activity relates to their 'holistic' authenticity" (2014, p. 1408) or their personality (Wang 1999).

Museums, therefore, are no longer the touchstones of objective authenticity. Many scholars have questioned the commodified role of museums in their initiative to orchestrate an authentic experience. Chhabra (2007) shares a negotiated framework, to address the muddled distortions of authenticity, generated by multiple and often conflicting ideologies. Authenticity in the context of pre-covid era dynamics, was being deciphered beyond 'an aura of rarification', offering a fluid/blended frame for hedonic spectacle (Chhabra 2007; Trant 1998). Aligned with this view, Prentice writes that the manner in which cultural products are showcased as cultural experiences demonstrates "emerging centrality of consumption as a contemporary hallmark of expression, rather than production (2001, p. 8). Collections today do not simply narrate the original work of an artist but they are emblematic of contemporary culture. Meanings of objectively authentic commodities and experiences are transformed into a cluster of markers tailored for targeted markets. Based on her study of curators of heritage museums in the State of Iowa (United States), Chhabra lend credence to the notion of negotiated authenticity:

- The curators attach highest significance to the objectively authentic ideologies while delivering messages in constructivist settings in their struggle to attract visitor markets;
- In face of their conflicting role, as most are reported as pursuing and defending the purest version of authenticity, a reflexively positioned negotiated stance is an ideal approach; and
- In view of the reviewed mission statements, it appears that the positions of museums are always negotiated in resonance with their political environment. Negotiation has been dictated by power relations which barricade pursuit of personal ideologies (2007, pp. 443–444).

In light of her findings, Chhabra proposes a negotiated model to address the backstage conflict. Her negotiation model calls for the toning of essentialist and constructivist ideologies to promote strategically mediated experiential cultural settings that facilitate social capital and active citizenship.

She opines that "this stance can reduce the dumbing of museums and foster hybrid authenticity and challenge the excess of fragmented and individualistic meanings facilitated by existentialism" (2007, p. 444). Her study results extend support for the need for experiential consumption to match the hedonistic and self-actualized pursuits of visitors.

Hede et al.'s (2014) three-person reference to museums also calls for hybrid authentic experiences. She refers to museums as the first person, the second person, and the third person. The first person denotes that a museum is an exposing agent. Environments are constructed using exhibitions and storytelling narratives that reveal plausibility of the museum as well that of the narratives presented (Pine & Gilmore 2007). Hede et al.'s second person refers to the museum visitor and the notion of the authentic self. It stresses on the processes through which consumers endorse themselves with authenticity. It is postulated that the extent of product involvement serves as a yardstick and influences audience expectations of authenticity (Grayson & Martinec 2004). In such scenarios, faults of authenticity are ignored. Visitors are expected to engage in experiential consumption and seek existentialist authenticity (an optimal state of mind and seeking truth of the moment experiences). Hede et al. also say that "visitors to museums come with a range of characteristics that influence their own authenticity, which we suggest will have a bearing on how authentic they perceive their visitor experience" (2014, p. 1398). In other words, an authentic state within the self shapes the manner authenticity is perceived and absorbed in an external setting. Last, materials are regarded as the third person. Materials help the first person describe the story while decoding is done by the second person (Bal 1996).

Clearly, the contemporary museums are actively engaged in authenticating visitor experiences through different touch points during the visitor voyage (Anton, Camarero, & Garrido 2018). Creative approaches are needed to transmit authenticity. Anton et al. note that exhibit and information overload at museums can overwhelm visitors and make them fatigued and less excited. Numerous reasons can be attributed to diminished attention such as satiation, physical weariness, competition, noise pollution, diversions, wrong choice, or weak/unattractive layout of the exhibit. Satiation implies that the more visitors consume something, the less they find it enjoyable. The consumed content is stated to cause satiation later in the post-visit stage. Therefore, it is important to identify the visit aspects that add to boredom and gather insights on optimal authentic experience stimulants that can help to reduce satiation or 'hedonic decay' (Anton et al. 2018).

Psychological and physiological processes, tourists go through, have received meager attention in documented literature. Anton et al. (2018) opine that route selected for the heritage voyage, length of time spent, and anticipation of the visit are key factors that can help to avoid or bolster satiation in visitors. It can be shrunk with the use of psychological mechanisms such as physiological feedback, change in perceptions and self-contemplation

(Galak & Redden 2015). Also, repetition can sway visitors in both positive as well as negative ways. For instance, it can either trigger boredom and block the mind or heighten curiosity in an object or artifact with opportunities to accumulate further knowledge (Steenkamp & Baumgartner 1992). Therefore, both, the museum exhibits and visitor experiences need innovative authenticating.

Anton et al.'s study shows that a planned route and content anticipation and length of time at the museum decrease affective responses and escalate satiation. That said, very short visits and decreased expectation can also have a reverse effect as "limited exposure time decreases consumer attention to brand information" (2018, p. 58). Therefore, it is crucial to understand the impact of various programs on visitors and prudently authenticate visitor routes. While keeping the core of the exhibits intact, the management and curators can amend authenticity to avert the state of hedonic decline during both online and onsite visits. Interactivity and level of connectedness can stimulate optimal authentic experience.

Anton et al. also offer a discussion of visitor satiation, using the optimal stimulation level theory. They postulate that based on the association between a person's emotional response and the stimulation received at an authenticated setting or from the authenticated self, the intermediate levels of stimulation are the most gratifying. Hence, exploratory behavior of visitors increases stimulation levels based on object and self-authentication. Examples include "curiosity, information seeking, variety seeking or innovative behavior" (Anton et al. 2018, p. 50). It is pointed out that crowding of exhibits at museums and excessive consumption triggers satiation, thereby implying that longer the time spent at the museum, more the satiation. Also, visits of short duration and prior-visit browsing of museum website cause satiation. In other words, no prior-knowledge and freedom to wander tends to decrease satiation. Type of museum visited also plays an important role along with physical weariness and mental fatigue. It is also reported that "longer visits and anticipating the content diminish visitors' emotional response. The level of attention might, however, be low if the visit is too short or if the content is not anticipated, either because individuals lack sufficient time or because they have insufficient knowledge to feel involved in and focused on the exhibition" (Anton et al. 2018, p. 58). Clearly, visitors need to be regarded as 'active emotional agents' and museums need to authenticate themselves innovatively to stimulate positive emotional responses, by offering existentialist experiences in objectified settings. For instance, personal interest and enthusiasm can be generated through online programming and by offering a mix of onsite experiences and 'immersive digital heritage programs', authenticated by using a variety of prompts and tangible markers (Kidd 2019, p. 55).

During the recent decades, online museum visitors have exponentially grown with growing demand for around-the-clock access (Marty 2007). Evidently, it has become crucial for museums to develop a sophisticated online pool of digital resource based on an innovative communication system. Marty (2007) conducted an online survey of visitors to nine online museums to gather insights on how visitors use the museum websites and how the digital museums are shaping their expectations/experiences. It is noted that the majority of the respondents are frequent visitors to the museum website looking for online visuals and research data. In fact, they are interested in taking online tour of galleries/interactive exhibits on a regular basis. Information, resources, and activities sought are similar to onsite visits and the majority of them use the website to get answers to their queries (such as timings, locale, and directions), watch artifacts, and stay updated on new exhibits.

Morrison, Lehto, and Day (2018) report a surge of digital nomads, people who employ technology to boost their nomadic lifestyles. With dwindling faith in traditional marketing, there is a need for museums to use reliable sources to narrate their story and reach the target markets. Trends are leaning toward influencer marketing strategies and the most trustworthy source of information is reported to be the users, thereby endorsing the continued need/growth of social media guides/stimulants (Talbot 2017). This calls for the use of networks such as Instagram, YouTube, Snapchart, and Digital Reality. Evidently, it is important for museums to get informed of latest developments and innovations in ICT. The next section shares initiatives of museums to embrace digitalization and scrutinizes the manner in which authenticity is negotiated on digital platforms.

Digitalization of museums

Museums are increasingly exploring opportunities to incorporate technology into their communications systems, "to do things differently", as more and more consumers embrace digitalization. Several benefits of Internet, for museums, are touted by Bowen and Filippini-Fantoni (2004) such as: (1) potential for global networking; (2) fast and convenient; (3) making inaccessible objects visible through virtual reality; (4) distributing information at less cost; (5) using Internet as a persuasive tool to transform online visitors to onsite ones; and (6) offering remote access to databases thereby facilitating research and scholarship. Furthermore, in support of crucial role of ICT in the museum sector, it can be argued that:

> An ever-growing part of our cultural heritage is stored in museum archives and repositories, where visitors can neither see nor access it. Such conditions contradict a museum's mission to "represent the world's natural and cultural commonwealth" (as stated in the AAM Code of Ethics for Museums). In their efforts to preserve cultural heritage, museums are also placing it under lock and key. But through the Web, a museum can now provide public access to its entire collection. Digital heritage projects have stored images of countless artifacts in

> collections databases. Despite such considerable efforts, few museums have fully embraced the diverse possibilities of having an unlimited space for display and communication. (Müller 2002, pp. 21–22)

As pointed out in Chapter 4, Information Communications Technology (ICT) has revolutionalized the communication system of museums. Museums are increasingly becoming storehouses and open spaces of knowledge instead of limiting themselves to repositories of tangible objects. An 'I am outstanding' model for effective tourism websites is proposed by Morrison (2009) encompassing fourteen key features:

- International – to communicate with global markets;
- Address travelers as individuals – personalizing the website experience;
- Monitor – check and evaluate in an ongoing manner;
- Outstanding – endeavor to be the best;
- Up-to-date – offer timely information;
- Target – identify and focus on different target markets;
- Social media – engaging conversations about the destination;
- Telephone-ready – making mobile phone applications available;
- Attractive – design attractive home pages;
- Network – partner with others for bigger success;
- Dynamic – build interactive digital platforms;
- Integrate – link all marketing communications;
- Niche markets – leverage special interest; and
- Great contents – offer comprehensive and useful information (Morrison et al. 2018).

Furthermore, demand has soared for user friendly and interactive exhibits. Museums can employ a variety of attributes to develop online exhibitions so that they can expand their outreach and span local, national, and global boundaries:

- *Space* – as online show creates a two-dimensional display, the social and physical presence is reduced due to the intimate interaction between the user and the monitor. Digital advanced viewing can challenge and expand our perception of art works. The virtual display removes objects from the referential frame of a traditional museum space. Virtual museum spaces can take on any shape they want, but they lack the conventional authority and emotion a museum building evokes;
- *Time* – online exhibitions are defined by the time that visitors need to access them. Web time is easily organized according to the user's individual needs;
- *Links* – surfing illuminates the language of the Web as a series of windows (or frames) and links. An online show speaks through a montage of images, sound, text, and design and the navigation of its pages.

Everything is just a click away. A visit to a virtual space might not be as intentional as a visit to the physical museum, where visitors dutifully wander through the galleries, even if they are not enchanted. Web semiotics encourage rapid decisions, and museums are challenged to make their voices heard within this new environment;

- *Storytelling* – only a small percentage of digitalized museum images are immediately compelling or engaging. Most digital reproductions only gain depth when they are presented as part of a larger story. Multimedia can lead to a diversity of voices in an exhibition, whether onsite or online. Storytelling creates a sense of space the Web deeply needs;
- *Interactivity* –online exhibitions must find ways of nurturing interactivity and facilitating access by including options such as: nonlinear but transparent navigation of informational resources; behind-the-scenes examinations of curatorial work; and open communication via email and guest books. Such increased access might lead to an open and interactive approach that permits visitors to become commentators, contributors, and even co-producers;
- *Production values* –the development of data standards and the digitalization of artifacts continue to be costly enterprises. Online curators do not have to worry about shipping, installation, conservation, or insurance issues. Working with digitalized information is cheaper, faster, and more flexible. And the low-production cost of online shows make them good tools for small and large museums to redefine and innovate themselves; and
- *Accessibility* –online accessibility is the main incentive of digital heritage programs. Online environments do not create limitations for people with disabilities (Müller 2002, p. 28).

Russo et al. (2008) argue that the practice of uploading and sharing of digital content, via social media applications, confirms and promotes the need for creative representation, search for identity, and cultural interactions. The authors share numerous success stories of social media projects such as the 'Sydney Observatory' by the Powerhouse Museum and the 'Educator Resource Center' by the National Design Museum. Medium used for these projects includes online videos, slide share, blogs, and a forum. On a positive note, social media has extended the museum role from information providers to self-exploration facilitators, offering opportunities to people to discern and interpret knowledge. It has enabled building or enhancement of popular knowledge-sharing systems enabling the cultural audience to share visuals, information, and experiences. The World Wide Web cannot be overlooked as it has a crucial role to play:

> It has become increasingly relevant to such core museum tasks as collecting, preserving, and exhibiting. Digitization of objects in digital

heritage programs has led to new forms of collection management and unparalleled access to virtual replicas of museum artifacts.

> This transformation is inspiring new forms of preserving and displaying of cultures both on- and off-line. Leonardo da Vinci's Last Supper is one of the great paintings of European art. Housed in the Santa Maria delle Grazie church in Milan, visitors can see it only in small groups, under tightly controlled conditions, which include a limited time-slot and prior reservations. At least, that's the theory. In today's world, the Last Supper has become a virtual painting, because few people ever really get to view it. (Preuss 2008, p. 21)

Digital heritage programs are becoming increasingly popular. Numerous other initiatives can be noted at the national level in different countries such as Australia and Canada. To cite an example, Australian Museums and Galleries On-line (AMOL) is a collaborative initiative between the Commonwealth, State/Territory governments, and the museum sector in Australia. Its National Directory of Museums showcases approximately 1000 museums and galleries and offers access to a database that includes 51 collections/400,000 items (National Library of Australia 2018). Department of Canadian Heritage has developed Canadian Heritage Information Network (CHIN) to offer access to unique collections to the museum community. It also offers access to millions of collections and helps the Canadian museums in documenting, managing, and sharing information about their collections (Department of Canadian Heritage n.d.).

Treinan (1993) stresses on the significance of context and the manner in which meaning of a tangible object is transmitted and communicated. This demand for 'connectedness and context' has enhanced the popularity of virtual museums. Schweibenz argues that the "virtual museums seek to describe the interrelated and interdisciplinary presentation of museum information with the help of integrated media. Connectedness is the quality that allows the 'virtual museum' to transcend the abilities of the traditional museum in presenting information and this does not merely mean to link objects together but to give visitors the opportunity to focus on their special interests by pursuing them in an interactive dialog with the museum" (1998, pp. 188–189). Müller (2002) has further inspected the relationship of museums with virtuality: (1) virtuality illuminates exhibition practices in museums; (2) it has opened a new path of communication and display to a broader audience and reproduces actual objects in digital formats; (3) helps to enhance 'curating online' strategies and comprehend the exhibits; and (4) artifacts and experiences can be simulated to make them enjoyable and engaging.

Joconde is a French museum database, maintained by the French Ministry of Culture. It encourages voluntary participation of museums to enrich the collection database and make it accessible. Besides catering to the informational needs of the general public, it is a useful resource for curators and managers in that it augments the cataloguing of collections and inventory. Such examples illustrate that digitalization has not compromised the objective authenticity of museums; instead, it has generated the ability to share and strategically negotiate object authenticity and make it meaningful in experiential and augmented/virtual settings.

Furthermore, augmented reality (AR) integrated with mobile technologies has become a useful tool. GPS-based AR can offer interactive and engaging tourism experiences (Dieck & Jung 2015). AR applications hold potential to provide interactive/enjoyable experiences and a level of comfort to tourists in unfamiliar settings. Museums can use this tool to educate visitors, about history, in an enjoyable manner. Kidd (2019) also writes that the virtual reality (VR) and AR applications enable museums to do things differently and creatively, especially by offering immersive digital heritage programs, to intensify emotional encounters with heritage. As an example, Table 9.1 illustrates

Table 9.1 With New Eyes I See

Participants were asked purposefully and vividly to empathize. Cyril (a botanist at the museum) became their conduit as they traced footsteps in a dramatic present tense to France in 1915 where he has wounded and later moved to Israel where he died. Participants moved from 'scene' to 'scene' at their own pace, taking time to feel – what he might have felt, or to situate themselves within the larger story about the War that was unfolding, as it were, around them (as Cyril might have himself might have done).
There were projections; old photographs, documents, animations. A voiceover. The sound of (Cyril's?) footsteps that could be followed between locations. There were silences; some poignant, some awkward. Found objects; a first aid box for participants to open and explore, a white lab coat upon which one of the animations was projected. Participants literally and metaphorically projected a new narrative within sites they often navigated in their everyday use of the city. It was a narrative that in subtle and not so subtle in ways intended to move participants to imagine themselves into the situation of this Other from the past.
Time-lapse footage of plants germinating, growing, and decaying was projected onto a polished memorial stone. In another moment a series of botanical images was pinned haphazardly to a tree alongside a telegram announcing Cyril's death. Curators made decisions about how best to bring what were clearly quite emotionally charged-events into proximity with the environment, and with the character development.
The tool for accessing the curated content was a mocked up military torch encasing the technology. It was hoped the simplicity of the interaction (triggered via RFID) would allow for heightened sociality of experience, avoiding the hardware becoming in any way a barrier to access or a distraction in the moment.
The participants huddled together around the map and together arrived at a consensus about how to way-find and where their exploration should begin. The RFID triggered the content to be delivered at each stop, and the curators had placed a number of other items for groups to interact with, whilst on their journey. There were, of course, an unlimited number other stimuli in the environment for participants to pull into the narrative and these became unanticipated arbitrators of experience.

Source: Kidd (2019, pp. 57–60)

how the 'With New Eyes I See' (WNEIS) project is designed to stimulate an emotional response. Kidd notes that, "the visitor is now often understood as very much an embodied and active agent, whether online, offline, or moving between the two (Drotner & Schrøder 2013; Kidd 2014; Parry 2013). While both augmented and virtual realities, in symbiosis, hold tremendous potential to facilitate immersive digital heritage experience (Woolford & Dunn 2013; Yoon, Elinich, Wang, Steinmeier, & Tucker 2012), Kidd writes that the most intriguing is the embrace of the seamless multimodal AR concept in museum environments. In Chapter 4, I illustrate this concept in a detailed manner.

The core of AR stimulates multiple sensory techniques and helps to design ways in which the virtual layer can respond to the AR prompts and offer a truly 'multisensory, personified/vibrant and physical museum experience' (Damala et al. 2016). The human-centered computing programs in museums "include a consideration for the body and 'the senses, the physical environment, and the social world' where visitors might be considered 'active agents in the process of interaction', embodied and situated" (Ciolfi 2015, p. 420). Centered on the human-based principles, the effort is to establish a symbiosis between artefacts, persons, and physical settings (Gottlieb 2008). In the WNEIS setting, the participants are able to negotiate their own route based on the information offered in the narrative by holding the torch and exploring with their hands. Kidd points out that "the theme of 'silence' has not been a real feature of the debate about digital possibility either within or beyond the museums and heritage scholarly community (Kidd 2019, p. 57). The author ponders if silence should be made "an active constituent within the narrative" and whether it may "play out within what was designed to be a social encounter" (Kidd 2019, p. 57). Table 9.1 gives a brief account of the WNEIS plan and design.

Three key points emerge from the WNEIS discussion: (1) ambiguity has potential and does not dispute a respectful and engrossing interpretation of a contested heritage; (2) WNEIS momentarily succeeded in changing the relationships between participants and the physical environments they traversed; (3) emphatic engagement, personified expressions, and silence were key triggers that helped reevaluate the previous suppositions associated with the existence of museums and their key purpose/value. The author calls for future research to identify voices, spaces, and other prompts that can perform the role of catalysts for empathetic engagement within animated and captivating events. The next section commences with a discussion on the significance of ICT in shaping the manner in which negotiated authenticity is being/can be authenticated and showcased.

Negotiated authenticity and its authentication in museums

Clearly, authenticity continues to be the focal point in the midst of a plethora of meaning attributions in both onsite and online museums (Chhabra 2007; Kidd 2014, 2019). The problematic manner, in which authenticity is presented and pursued in onsite and online museums, is evident in the example of 'Seeing it for Real or Going to See for Real Life' experience presented, by the People's Museum of History in Manchester, for the school children. The show motivated the junior school children at the museum and the experience inspired them, particularly those aspects that enabled them to 'see it for real', that is, creatively engage with the live performance of the actors in costume. Jackson and Leahy write that:

> the children attributed authenticity to all of their experiences in the museum, taking their cue from the presence of real objects and not differentiating between the diverse forms of interpretation (including theatre) that they encountered and, significantly, performed (e.g. playing dressing shop and dressing up). The desire to apprehend the museum as a repository of truthfulness may have been partly fueled by the children's enthusiasm for an escape from the classroom into this space of objecthood and external authority. In effect, their belief in the authenticity of the disparate elements of their museum experience – encompassing taxonomic displays, room sets, handling and role-play, interpretation, performance – has evoked the role of the museum as a source of legitimacy and truth. The novelty of the museum heightened the children's receptivity and predisposed them to regard what was clearly make-believe – the theatrical performance – as another manifestation of the museum's authority (2005, pp. 321–322).

Using different types of reactions, Jackson and Leahy (2005) conclude that both affective and cognitive experience/learning assist in enhancing learning in informal settings such as a museum. One might ask: what kind of authenticity does digital heritage promote and how can it be authenticated? Müller argues that digital reproductions of museum exhibits have both pros and cons. For instance, the digital showcasing of museum objects can prove to be asset. However, the valued social and civic setting within which the experience happens becomes absent. For the most art, the digital reproduction presents the visualness or acoustic aspects of the object and these do not offer a sense of touch as the objects cannot physically held. In fact, "virtual programs eliminate the physical dimension altogether as well as the momentum created by the object's physical presence (after all, bytes have no aura). But the digital copy can offer new venues for contextualizing the object and investigating its informational layers as well as interactive options for exploring its characteristics and history" (Müller 2002, p. 24).

Numerus contemporary museums are offering story-centered exhibitions of objects, while still retaining/advocating the authenticity of the actual physical experience. This explains why reproductions are rarely showcased. Technology and media tools are often leveraged to augment the visitor experience. For instance, lighting can sensationalize the charisma of an object, stories can be narrated through aural tours, and historical synopsis can be offered via videos. All these tools are valuable in offering good quality experience. Likewise, the quality of presentation of the digital counterpart can augment audience understanding and recollection of its history. From a positive standpoint, Müller writes that:

> Removed from a physical space, however, the digital experience might encourage a more rational reception of the artifact on display. Today's debates on virtuality recall those on art and reproduction technologies that took place in the first part of the 20th century. As early as 1934, philosopher Walter Benjamin wrote in The Work of Art in the Age of Mechanical Reproduction: The technique of reproduction detaches the reproduced object from the domain of tradition. By making reproductions, it substitutes a plurality of copies for a unique existence. Benjamin's dictum, that art looses its aura and immediacy of experience through the possibility of its mechanical reproduction (its reproducibility), was written during the newly emerging development of mass culture, (e.g. film), and at the time of radical attacks on "authentic art" by artists themselves, as in Dada or Surrealism (2008, p. 24).

Negotiated authenticity can also be appropriated through shared practices. Museums are institutions that project shared identity through 'collective memory making' practices (Zhang, Xiao, Morgan, & Ly 2018, p. 117). For instance, they are 'material testimonies of national identity' as they build domineering/preferred national identities (Zhang et al. 2018, p. 116). Tangible elements shape memories; however, meaning and imagination are equally important in offering interactive and meaningful museum experiences (Park 2011). Museums transmit preferred meaning of national identities despite their claim against support of disputed identities (Feldman 2012). Traces of negotiated authenticity, authenticated with the use of selected themes, is evident here.

Next, several scholars have questioned the authenticity of reproductions and their value in helping understand heritage and history of a place. In reply, recent studies have offered a discourse on the manner negotiated authentication guides the creation/presentation of heritage resources, both online and onsite. A digital reproduction does not necessarily diminish the value of a craft or an artistic act because since early years, negotiated reproductions have been used to inform public of history and culture. As an example, our knowledge of Greek culture mostly comes from reproductions produced by Rome (Müller 2002). Photography and films, which are also examples of reproduction technologies, are considered a form of art today. Müller notes that: "a mix of original and reproduced works inform much of our current knowledge of cultural history. The moment of intense encounter with a work of art or a historical artifact can have a long-lasting influence on our emotional curiosity and our quest for knowledge. By often, love for art and culture is nurtured by reproductions: book and catalogue illustrations, postcards; posters; and now thumbnails. While the singularity of the artifact fades in its duplication, most of its informational layers stay intact. Education and understanding of culture is based on this information and not exclusively on the emphatic experience of the objects' presence" (2002, p. 27).

Closing comments

The deschooling of museums has opened doors and offered public access to invisible repositories. Museums are listening to their visitors and adapting innovative approaches to authenticate their resources while bargaining to stay objectively authentic to their core. Today, authenticated narratives are using a praxis approach to stimulate reflection and immersive experiences, both onsite and in digital environments, forging a connection between the "then" and "now" (Graburn 1998). Immersive digital heritage programs hold potential to evoke emotional reactions and heighten visitor engagement with heritage (Kidd 2019). For instance, 'With New Eyes I See' explores the negotiated intersection between silence, empathy and embodiment. Presentation and selection of digital objects is often shaped by the transformative cultural-politics of authentication. Returning to the core question: how do the above discourses shape the notion of negotiated authenticity and how are objects and experiences authenticated by museums? In fact, most studies have overlooked important insights associated with the journey through which negotiated authenticity is bestowed or authorized.

Zhang et al. (2018) use collective memory theory to demonstrate processes pursued by different stakeholders of heritage tourism to reconstruct national identity through museums. The authors offer a discourse on how identity (reconstruction) is pursued in two contesting museums: the Macao Museum and the Hong Kong Museum. Both museums are part of China's nation building efforts; however, struggle with the perceptions of 'Chineseness' have made the museums in Hong Kong and Macao highly contested with regard to orchestration of evidence to support their stance on national identity (Wang & Law 2017). Zhang et al. argue that museums negotiate their disputed national identities by authenticating preferred ideologies.

We cannot deny the role of museums as authenticating agents of culture and heritage. It is this very role that defines and establishes their authority. Museums are hidden agents through which authenticity is represented; in fact, authenticity is created in and by them. Varutti (2018) stresses on scrutinizing the authentication of heritage objects by museums rather than dream pursuit of an unadapted piece of heritage or become fixated with a particular version of object authenticity. He points out that authentication is embedded in a large number of museological practices such as in the commission and purchase of newly made traditional objects. The acquisition function of the museum, in fact, can be questioned as it certifies cultural experts and endorses that the conventional methods have been employed by the maker. In doing so, the museum is confirming a specific degree of authenticity in the submitted or showcased object. Although copying and imitation practices are key to the cultural production and reproduction processes in museums, they can be controversial from an objective authenticity standpoint.

Varutti further writes that "in the light of this tension between the significance of the reproduction of cultural artifacts for today's indigenous groups, and the inherent emphasis of museums on the 'original', it seems interesting to consider the role that museums can play in processes of authentication and in particular, in the authentication of an indigenous cultural heritage that is essentially based on the reproduction of the worlds artworks" (2018, p. 43). Although Varutti discusses the role of museums as authenticating agents in the context of indigenous heritage, his views are applicable to all types of heritages and cultural displays.

Authenticity needs to be completely deciphered as a process, rather than a static notion or an intrinsic trait of heritage objects and museums play a key role in negotiating/channeling the authenticating process (Varutti 2018). Authentication is beneficial in that: ethnographic museums, for instance, can become instrumental to the revival of indigenous cultures through the validation and authentication of contemporary indigenous artefacts. Such authentication enables a shift from the category of 'items of indigenous art and craft' to the much more salient and politically powerful category of 'items of indigenous cultural heritage' (Varutti 2018, p. 44). Authenticity is often appropriated through the process of classification and cataloging. Different elements are selected to signal authenticity. The authentication process in museums is complex because:

> It usually aims to probe the origins of an artefact in order to detect forgery, illicit traffic or other problematic situations. To this end, authentication generally involves a set of actions focusing on the authorship, materiality, techniques of production, and ownership of an artefact, among other considerationions. These actions may include – for instance – laboratory examination of the materials, analysis of condition reports and legal documentation where available, research on provenance, analysis of similar objects in museum collections through digital visual analysis tools and so on.
>
> Conversely, if we consider what kind of museum practices index and create authenticity, we can see that the acquisition and commission of heritage objects encompass and engender several museum practices that are part and parcel of the making of authenticity – or authentication. Cultural authentication can also be enacted through specific display solutions, using ad hoc juxtapositions among objects. What is being authenticated through museum acquisition and commission is not only the object, but also its maker, as a producer of authentic indigenous cultural heritage, with appropriate indigenous and cultural credentials. (Varutti 2018, p. 51)

However, the new museum practices, where newly made objects are commissioned and endorsed for traditional authenticity, demonstrate that museums are playing an influential authenticating function through strategically selected/preferred authentication procedures. According to Bruner, "where authenticity is certified by authoritative actors, the issue of authenticity merges into the notion of authority" (1994, p. 400). A critical discourse is required to scrutinize the authority exercised by museums to shape preferred narratives and objects for selective showcasing. Limited studies have examined the role of museums as authenticating agents of the objects and heritage experiences. Literature is sparse on scrutiny of digital authentication in the context of museums.

Furthermore, COVID-19 has presented a unique set of challenges to the museums across the globe. Museum professionals face the challenge of remaining open even during the closure of their physical spaces. The pandemic has raised the question of their relevance. According to Kahn, "if museums take the 'let's sit this out, and see what happens approach' they are far less likely to emerge with evolved, healthy and flexible identities that will be needed to continue their roles as preservers of knowledge and transmitters of communication in the post-epidemic world. Online or digital exhibitions, which replicate, extend and supplement physical collections are gradually becoming more commonplace, but among museum professionals, there is a general lack of consensus about what a digital or online exhibition might be" (2020, p. 4).

As museums start to delve more deeply with the realities of the new museology in the wake of the pandemic, early initiatives on what shape a digital museum can take conflict with their original purpose. Kahn further points out that, "at the heart of this tension lies the need to find ways to highlight and bridge two of the most important roles museums have: as sources and spaces for scholarly communication (in other words, museums as knowledge repositories) and as places for science communication (museums as sources of informative entertainment). Both of these roles have, until now, been situated in the physical space of the museum. However, in thinking through an evolution which is digital and hybrid, COVID-19 offers the potential for museums to reveal their inner workings, and potentially share much more with their virtual visitors" (2020, p. 4).

But on a positive note, museums have always evolved and embraced a broader public health role even though, in the intra-pandemic times, they have been mostly restricted to digital platforms. They are weathering the COVID-19 storm in an innovative manner. As an example, Agostino, Arnaboldi, and Lampis (2020), in their study of 100 Italian State museums, report that closure of physical sites during lockdown did not disrupt their initiatives. Instead, a sharp rise in online activity has been noted, in terms of sharing online cultural material and social media initiatives. It is not only the museums that face the challenge of remaining open even during closure of their physical spaces, ethnic restaurants are also struggling to remain in business. How ethnic cuisines transmit cultural messages and to what extent they have retained their authenticity also form an important field of inquiry. The next chapter situates the authenticity and authentication discourse in the context of ethnic restaurants.

References

Agostino, D., Arnaboldi, M., & Lampis, A. (2020). Italian state museums during the COVID-19 crisis: From onsite closure to online openness. *Museum Management and Curatorship*, 1–11.

Anton, C., Camarero, C., & Garrido, M. (2018). A journey through the museum: Visit factors that prevent or further visitor satiation. *Annals of Tourism Research*, *73*, 48–61.

Bal, M. (1996). *Double exposures: The subject of cultural analysis*, London: Routledge.

Belk, R. (2001). The role of possessions in constructing and maintaining a sense of past. *Advances in Consumer Research*, *17*(1), 669–676.

Beverland, M., Lindgreen, A., & Vink, M. (2008). Projecting authenticity through advertising. *Journal of Advertising*, *37*(1), 5–15.

Bowen, J. P., & Filippini-Fantoni, S. (2004, March). Personalization and the web from a museum perspective. In *Museums and the web* (Vol. 4). Toronto, Canada: Archives & Museum Informatics.

Britton, S. (1991). Tourism, capital, and place: Towards a critical geography. *Environment and Planning*, *D*(9), 451–478.

Bruner, E. (1994). Abraham Lincoln as authentic reproduction: A critique of post-modernism. *American Anthropologist*, *96*(2), 397–415.

Burcaw, G. (1975). *Introduction to museum work*. Nashville, TN: American Association for State and Local History.

Chhabra, D. (2007). Exploring market influences on curator perceptions of authenticity. *Journal of Heritage Tourism*, *2*(2), 110–119.

Chhabra, D. (2010). *Sustainable marketing of cultural and heritage tourism*. Routledge.

Ciolfi, L. (2015). Embodiment and place experience in heritage technology design. In Michelle Henning (Ed.), *The international handbook of museum studies: Museum media* (pp. 419–446). Hoboken, NJ: Wiley.

Damala, A., Vaart, M., Loraine, C., Hornecker, E., Avram, G., Kockelkorn, H., & Ruthven, I. (2016). Evaluating tangible and multisensory museum visiting experiences: Lessons learned from the meSch Project. Accessed Janaury 22. http://mw2016.museumsandtheweb.com/paper/evaluating-tangible-and-multisensory-museumvisiting-experiences-lessons-learned-from-the-mesch-project/.

Department of Canadian Heritage. (n.d.). CHIN. Retrieved from https://www.canada.ca/en/heritage-information-network.html

Dieck, C., & Jung, T. (2015). A Theoretical Model of Mobile Augmented Reality Acceptance in Urban Heritage Tourism. Current Issues in Tourism.

Drotner, K., & Schrøder, K. C. (2013). *Museum communication and social media: The connected museum*. New York, NY: Routledge.

Falk, J. (2006). An identity-centered approach to understanding museum learning. *Curator: The Museum Journal*, *49*(2), 151–166.

Feldman, J. (2012). Exhibiting conflict: History and politics at the museo de la memoria de ANFASEP in Ayacucho, Peru. *Anthropological Quarterly*, *85*, 487–518.

Galak, J., & Redden, J. (2015). The properties and antecedents of hedonic decline. *Annual Review of Psychology*, *69*, 1–25.

Gottlieb, H. (2008). Interactive adventures. In Loic Tallon (Ed.), *Digital technologies and the museum experience: Handheld guides and other media* (pp. 167–178). Lanham, MD: AltaMira Press.

Graburn, N. (n.d.). The "Full Service" Museum. Berkeley: Manuscript

Graburn, N. (1998). A quest for identity. *Museum International*, *50*(3), 13–18.

Grayson, K., & Martinec, R. (2004). Consumer perceptions of iconicity and indexicality and their influence on assessments of authentic market offerings. *Journal of Consumer Research*, *31*, 296–312.

Harrison, J. D. (2005). Ideas of museums in the 1990s. *Heritage, museums and galleries: An introductory reader*, 38–53.

Heard Museum (n.d.). Mission Statement. Retrieved on October 1, 2019, from https://heard.org/home/wp-content/uploads/2016/02/Heard-Museum-Group-Tour-Information-2017-2018.pdf.

Hede, A. M., Garma, R., Josiassen, A., & Thyne, M. (2014). Perceived authenticity of the visitor experience in museums: Conceptualization and initial empirical findings. *European Journal of Marketing*, *48*(7/8), 1395–1412.

Hein, G. (1998). *Learning in the museum*. London: Routledge.

Hermitage Museum (2019). About Us. Retrieved on September 20, 2019, from https://hermitage.nl/en/about-museum/our-mission/.

Hooper-Greenhill, E. (2000). *Museums and the interpretation of visual culture*. London: Routledge.

HyungYu, P. (2011). Shared national memory as intangible heritage: Reimagining two Koreas as one nation. *Annals of Tourism Research*, *38*, 520–539.

Jackson, A., & Leahy, H. R. (2005). 'Seeing it for real ...?'—Authenticity, theatre and learning in museums. *Research in Drama Education*, *10*(3), 303–325.

Jones, S. (2010). Negotiating authentic objects and authentic selves: Beyond the deconstruction of authenticity. *Journal of Material Culture*, *15*(2), 181–203.

Kahn, R. (2020). Locked down not locked out–assessing the digital response of museums to COVID-19. *Impact of Social Sciences Blog*.

Kelly, L., Cook, C., & Gordon, P. (2006). Building relationships through communities of practice: Museums and indigenous people. *Curator: The Museum Journal*, *49*(2), 217–234.

Kidd, J. (2014). *Museums in the new mediascape: Transmedia, transmedia, ethics*. Farnham: Routledge.

Kidd, J. (2019). With New Eyes I See: Embodiment, empathy and silence in digital heritage interpretation. *International Journal of Heritage Studies*, *25*(1), 54–66.

MacDonald, G., & Alsford, S. (1995). Museums and theme parks: Worlds in collision? *Museum Management and Curatorship*, *14*(2), 129–147.

Marty, P. F. (2007). The changing nature of information work in museums. *Journal of the American Society for Information Science and Technology*, *58*(1), 97–107.

Metropolitan Museum of Art (2019). Mission Statement. Retrieved on September 20, 2019, from https://www.metmuseum.org/-/media/files/about-the-met/annual-reports/2014-2015/annual-report-2015-mission-statement.pdf.

Morrison, A. (2009). *Hospitality and travel marketing* (4th ed.). Boston, MA: Cengage Learning.

Morrison, A., Lehto, X., & Day, J. (2018). The Tourism System, 8th ed. Dubuque, IA: Kendal-Hunt. National Library of Australia. AMOL. Retrieved on November 13, 2018, from https://www.nla.gov.au/pathways/jnls/newsite/view/396.html.

Müller, K. (2002). Museums and virtuality. *Curator: The Museum Journal*, *45*(1), 21–33.

National Library of Australia (2018). Report. Retrieved from https://www.nla.gov.au/corporate-documents/annual-reports.

National Museum of Scotland (2019). Mission Statement. Retrieved on October 2, 2019, from https://www.nms.ac.uk/about-us/our-organisation/vision-and-values/.

Parry, R. (Ed.). (2013). *Museums in a digital age*. Routledge.

Pine, B., & Gilmore, J. (2007). Museums & authenticity. *Museum News: Washington*, *86*(3), 76.

Pine, B. J., & Gilmore, J. H. (2007), *Authenticty: What consumers really want*. OH: Strategic Horizons LLP.

Prentice, R. (2001). Experiential cultural tourism: Museums & the marketing of the new romanticism of evoked authenticity. *Museum Management and curatorship*, *19*(1), 5–26.

Preuss, D. (2008). Body worlds: Looking back and looking ahead. *Annals of Anatomy-Anatomischer Anzeiger*, *190*(1), 23–32.

Russo, A., Watkins, J., Kelly, L., & Chan, S. (2008). Participatory communication with social media. *Curator: The Museum Journal*, *51*(1), 21–31.

Schweibenz, W. (1998). The "Virtual Museum": New Perspectives for Museums to Present Objects and Information Using the Internet as a Knowledge Base and Communication System. *Isi*, *34*, 185–200.

Silverman, L. (1995). Visitor meaning making in museums for a new age. *Curator: The Museum Journal 38*(3), 161–169.

Steenkamp, J.-B. E. M., & Baumgartner, H. (1992). The role of optimum stimulation level in exploratory consumer behavior. *Journal of Consumer Research*, *19*(3), 434–448.

Stephen, A. (2001). The contemporary museum and leisure: Recreation as a museum function. *Museum Management and Curatorship*, *19*(3), 297–308.

Talbot, K. (2017). Influencer Marketing Strategy. Retrieved on November 13, 2018, from https://www.forbes.com/sites/katetalbot/2017/12/22/how-to-nail-your-2018-influencer-marketing-strategy/#186eace624bf.

Te Papa Museum (2019). Statement of Intent. Retrieved on October 5, 2019, from https://www.tepapa.govt.nz/sites/default/files/tp_statement_of_intent_2017-2021_online_002.pdf.

Trant, J. (1998). When all you've got is "The Real Thing": Museums and authenticity in the networked world. *Archives and Museum Informatics*, *12*(2), 107–125.

Treinan, H. (1993). What Does the Visitor Want from a Museum (Sch 1998 paper).

Varutti, M. (2018). 'Authentic reproductions': Museum collection practices as authentication. *Museum Management and Curatorship*, *33*(1), 42–56.

Wang, L., & Law, R. (2017). Identity reconstruction and post-colonialism. *Annals of Tourism Research*, *63*, 203–204.

Wang, N. (1999). Rethinking authenticity in tourism experience. *Annals of Tourism Research*, *26*(2), 349–370.

Watkins, C., & Mortimore, P. (1999). Pedagogy: What do we know. *Understanding pedagogy and its impact on learning*, 1–19.

Witcomb, A. (1999). Museums as cultural brokers: Producing rather than representing communities. In B. Henson (Ed.), *Exploring culture and community for the twenty-first century: Global arts link: A new model for public art museums* (pp. 101–104). Ipswich, Queensland: Global Arts Link.

Woolford, K., & Dunn, S. (2013). Experimental archaeology and games: Challenges of inhabiting virtual heritage. *Journal of Computing and Cultural Heritage*, *6*(4), 16.

Yoon, S., Elinich, K., Wang, J., Steinmeier, C., & Tucker, S. (2012). Using augmented reality and knowledge-building scaffolds to improve learning in a science museum. *Computer-Supported Collaborative Learning*, *7*, 519–541.

Zhang, C., Xiao, H., Morgan, N., & Ly, T. (2018). Politics of memories: Identity construction in museums. *Annals of Tourism Research*, *73*, 116–130.

10 Negotiated authentication of ethnic cuisines

This chapter examines 'negotiated authenticity (or otherness)' of ethnic cuisines, authenticity perspectives of the food heritage concept, and authentication process employed to authenticate 'othered' cuisines. It also discusses the future of ethnic cuisines during the intra- and post-pandemic times.

Food heritage is an important element of heritage tourism, especially ethnic tourism. Ethnic tourism can be defined as a type of tourism that is inspired by the quest to seek novel cultural experiences such as "visiting ethnic villages, ethnic theme parks, shopping for ethnic crafts and souvenirs and activities focused on tasting exotic and novel foods" (Lai, Lu, & Liu 2019; Yang, Wall, & Smith 2008). Extant literature acknowledges that food functions as an authenticating agent for ethnic experiences (Barbas 2003; Chhabra, Lee, Zhao, & Scott 2013; Robinson & Clifford 2012; Sims 2009). Moreover, it is an important mechanism and provides an ideal setting for accumulating cultural capital (Anderssons & Mossberg 2004). This chapter extends the discourse on authenticity in the context of ethnic cuisines by examining how it is negotiated and depicted through selected markers.

Negotiated "othered" food offerings are sought by consumers (Lai et al. 2019) and in response, restaurants have endeavored to authenticate authenticity of their cuisines in a manner that appeals to their patrons (Chhabra et al. 2013; Stock & Schmiz 2019). Several studies illustrate that most patrons prefer to experience the "other" in an objectively authentic setting but in a manner conducive and pleasurable to their lifestyle (Lai et al. 2019; Song, Phan, & Kim 2019). This chapter examines 'negotiated authenticity (or otherness)' of ethnic cuisines, authenticity perspectives of the food heritage concept, and authentication process employed to authenticate 'othered' cuisines. Toward the end, it discusses the future of ethnic cuisines during the intra- and post-pandemic times. Key topics of scrutiny are:

- Food as a promoter of ethnic heritage
- Negotiated authenticity of ethnic cuisines and authentic atmospherics
- Popular authenticating markers and authentication process of ethnic cuisines
- Future of ethnic cuisines in the post-pandemic times

Food is an essential component of ethnic tourism because it adds to the experiential value of a destination by offering immersed opportunities. According to Long (1998), food is an important way of soaking into another culture; it offers a path to experience the 'other' at a sensory level. Game (1991) says that an important feature of travel is to eat local food from the 'other' culture. Local/traditional food complements tourist experience at a destination (Morrison, Lehto, & Day 2018). Traditional food showcases culture, history, and lifestyle of a local community (Chhabra, Zhao, Lee, & Okamoto 2012) in terms of ingredients, preparation techniques, utensils, or the manner food is served (Shariff, Mokhtar, & Zakaria 2008). Its authentic elements are often drawn from the wisdom and proficiency of local people (Omar, Mohd Adzahan, Mohd Gazali, Karim, Abdul Halim et al. 2011).

Ethnic heritage tourism holds potential to serve as a preservation agent of ethnic food and ethnic restaurants have impacted the manner in which cultural food knowledge is imparted to the patrons. Governments of host countries are keen on supporting ethnic tourism to promote local traditions, cultural diversity, and economic benefits. Kuhn, Haselmair, Pirker, and Vogl postulate that "when tourists visit locations in which ethnic tourism has been established on the basis of the migrants' traditions and culture, they can have 'exotic' and unfamiliar experiences within their own country" (2018, p. 2). Food holds immense appeal among heritage visitors because it is one of the fundamental cultural expressions of historical narratives, rituals, and festivals (Horng & Tsai 2012). In the same vein, several authors have shared pitfalls associated with commodification and misrepresentation of heritage in an effort to make it appealing and give it a competitive edge (Mintz 1996; Okumus, Okumus, & McKercher 2007; Stock & Schmiz 2019). However, on a positive note, global factors can shape a nation's food culture but they do not necessarily marginalize or erode the localization effects of food culture; both can co-occur and converge (Henderson 2007).

The notion of food heritage stresses on the most traditional form of food growth and production (Ramli, Zahari, Ishak, & Sharif 2014). It relates to the cultural distinctiveness of a person, community, or region and refers to agricultural practices by rooting a community to its place and in a historical/traditional context (McCoy 2012). The underlying tenet is locally and naturally grown food with traditional production methods and unaltered tastes (Wahid, Mohamed, & Sirat 2009). Ramli et al. state that, "food heritage can be reflected from the environment history, belief, ideology and food technology of society in an era or period of time" (2014, p. 408). Food heritage can be delineated into two categories: common and extinct. The first category refers to the food consumed in everyday lives, whereas the second category refers to food that is slowly fading out or becoming defunct (Ramli et al. 2014). Five core elements define the concept of food heritage and these hold mirror to objective and negotiated authenticity versions of authenticity: historical traits (such as habit forwarded from the past to the present generations, natural, original, and localized), food attributes, distinctive value, utility, and its facet of assimilation (Guerrero, Guàrdia, Xicola, Verbeke, Vanhonacker, et al. 2009; Ramli et al. 2014).

It cannot be denied that authentic showcasing of food is often shaped by social and political interests. For instance, ethnic foodways "as specific zones, spatial and social, based upon a cultural tradition and its food", (Chaney & Ryan 2012, p. 311; Stock & Schmiz 2019) can segregate communities as these are often unevenly distributed/showcased stocks of cultural capital (Naccarato & LeBesco 2013). Stock and Schmiz (2019) scrutinize the role of ethnic entrepreneurs who use marketing strategies "to actively stage authenticity and create new tastes in their ethically marketed eateries" (2019, p. 1). In the transformation process, for instance in regard to the ethnic cuisine in Berlin, outcome is hybrid where the original elements are adapted to resonate with the local trends and preferred flavors. Self-positioning of an entrepreneur can vary, based on social and economic interests, for different target markets which in turn shape the manner traditional food is authenticated or gentrified. As lines between cultural capital and economic exploitation become blurred, preferred markets are selected or excluded on grounds of autonomy and authority. While scrutinizing ethnic cuisines of Singapore, Henderson reports:

> traditionally based foods seem to be too deeply embedded in Singapore society and culture to disappear, yet they will continue to mutate and a national cuisine perhaps remains a work in progress akin to the wider story of the city/state. Food and dining in present-say Singapore in conformity with much of the rest of the world, has been shaped by the flows of the capital, people, goods and ideas which characterize globalization. The result is a degree of internationalization, but this does not necessarily equate with homogenization. To label a cuisine local and in opposition to the global is perhaps simplistic, suggesting that the globalization-localisation and homogeneous-heterogeneous dichotomies may not be the best approach to conceptualizing food cultures and their evolution. Food is a link with the past, representing continuity and familiarity, and inspires a yearning for what are perceived to be authentic tastes and experiences on the part of residents. Visitors may share such longings and, for them, food can be a means of connecting to and learning about a destination (2012, p. 913).

Undoubtedly, food histories are nested within a romanticized and often illusionary dialectic prompted by local, cultural, political and economic interests (Pratt 2007). There is a recognition that food has its own real or pseudo/hyperreal heritage which deserves to be conserved (Henderson 2012; Stiles, Altıok, & Bell 2011). Henderson (2012) identifies three key discourses associated with food culture in present-day environments: (1) relationships shaped by the global-local nexus, (2) food as a symbol of identities at the national and ethnic level, and (3) notion/conservation of food heritage. While the first two perspectives reinforce a negotiated stance, the third perspective calls for food culture to be grounded in objectively authentic ideologies.

Authenticity and the authentication process

Literature on authenticity of ethnic cuisines has soared in recent years, offering both demand and supply perspectives. By furthering the discourse to authenticating agents and authentication of ethnic cuisines, some authors have unlocked new research paths. Different attributes of authenticity are reported that align with the purest version or the hybrid notion of authenticity. An earlier study by Lu and Fine (1995) shares dialogical efforts by managers of Chinese restaurants in their desire to freeze desirable part of the traditions while embracing change. Mohammad and Chan (2011) examine the perception, belief, and values of authentic cuisine in ethnic restaurants in Malaysia. The authors report four core attributes of food authenticity: heritage and consistent style, recalling (aromatic), uncontaminated, and harmonious. Cohen (2007) equates authenticity to genuineness in terms of traditional practice, repeat patronage of an intact/perfect product, and sincere associations. Local patrons seek authentic food taste and the manner in which the food is prepared. It is argued that the authentic spirit of production should remain intact through appropriate resources and initiatives. Mohamad and Chan say that, "the role of local customers should not be ignored in determining the authenticity within a particular social context (food, restaurants, and dining). If food is changed to appeal to the taste of foreign tourists, then traditional preparation/ingredients of the region can be lost. And this modification can have broader implications on cultural sustainability (Chappel 2001). Hence, it is important to ensure that the authentic cuisine of a region and marketable local regional foods are approached with a delicate balance" (2011, p. 464).

Authenticity of heritage products can mirror the producers' spiritual devotion to consistent style and method of production. Producers should be willing to dedicate time and effort to produce by traditional methods to keep the essence of authentic production (Liao & Ma 2009). Research reports from pre-pandemic times suggest that authenticity of ethnic food is not a stand-alone draw for target markets. Complete and aesthetic dining experience has to be offered which is determined by the physical environment and other intangible markers such as language and the traditional way of greeting. For instance, Liu, Li, DiPietro, and Levitt (2018) stress on the importance of authentic atmospherics besides food authenticity. Customers who patronize ethnic restaurants mostly seek an authentic dining experience that can be categorized into three domains: atmospherics, food ingredients, and the manner in which it is served. The authors make several suggestions:

- Managers in mainstream ethnic restaurants should ensure that the dining environment showcases the ethnic culture through cuisine name and presentation, atmospherics and the restaurant interior and exterior, and engrossing stories about the ethnic culture and history of the cuisine/restaurant;

- Maintaining food authenticity can augment customer's favorable evaluation of food quality and further increase its perceived value. Managers should, therefore, use food authenticity as a benchmark of quality control;
- Cultural familiarity and cultural motivation can optimize dining experience. Therefore, mainstream ethnic restaurants should strive to offer authentic atmospherics and food; and
- Perceived quality significantly influences perceived value, and managers of mainstream ethnic restaurants should move beyond leveraging cultural authenticity to gain competitive advantage and rather focus on food quality (Liu et al. 2018).

Jang, Liu, and Namkung (2011) scrutinize the effects of authentic atmospherics in the context of Chinese restaurants in the USA. They illustrate that authentic atmospherics are deliberately designed spaces to trigger desired emotional responses that can influence purchase intentions of target markets. Restaurant studies often relate to atmospherics in the context of service quality and authentic atmospheric indicators refer to "ethnic art, décor, external façade such as furnishings, painting, table setting, name and various stereotyped signals that generate an ethnic experience and ambience of a particular culture" (Jang et al. 2011, p. 667). Also, it can be argued that atmospherics shape consumer perceptions of authenticity. Furthermore, Tsai and Lu (2012) explain that authentic ambience of a restaurant is a key indicator of a satisfactory experience.

Jang et al. report a positive and significant effect of authentic atmospherics on emotions and find that evoked positive emotions are a significant intervening factor between atmospherics and behavioral intentions. Jang et al. stress on negotiated authentication of ethnic authenticity by offering "sophisticated cultural cues along with factors such as quality authentic food, good service, and appropriate prices" (2011, p. 675). The authors offer several suggestions to the managers of Chinese restaurants in the USA, in terms of authenticating markers and evocation of positive emotions. Penchant for authentic décor and food is regarded as the core influencing point that shapes guest's satisfactory dining experience and perceptions of authenticity. In other words, both atmospherics and cuisine need to be authenticated using objective authenticity as a reference point, to boost repeat patronage of loyal markets.

Moving forward, ethnic restaurants often serve as cultural ambassadors abroad in that they convey a 'sense of otherness' and foreign heritage to local patrons who quest for ethnic flavors (Jang et al. 2011). They are promoted with objectively authentic culture messages, which are especially convincing if the owner belongs to that culture. The traditional food and atmospherics then become a window of cultural expression. Ramli et al. (2014) reiterate the necessity of strategic cross-culturing of ethnic cuisines because of acculturation and assimilation practices that generate the need for a distinct 'othered' food identity while, at the same time, commodifying to appeal to local markets in migrating regions. The promotional messages of one local and one ethnic restaurant from Europe illustrate this point. The Walliser Keller restaurant in Zurich (Switzerland) promises a unique and authentic experience negotiated to match the need/taste of each customer by offering an opportunity to cook it in the desired manner:

> **Welcome to the Walliser Keller: "The Swiss Restaurant" since 1959**
>
> To us, "Swiss Alpine cuisine" means bringing a portion of the "Swiss mountains" to you at lunchtime or in the evening. The rustic cellar vault brings to mind a typical wine cellar in the vineyards of the Valais. We put the flavor of the Swiss Alps on your plates;
>
> **Experience an amazing cheese spectacle**
>
> The original cheese fondue comes to you in eight different variations, as dictated by Valais tastes. Among them, you will also find a vegan and a non-alcohol version of the cheesy spectacle. And when your meat is brought to you on a hot stone, cook it exactly as you want it. Of course, our menu also includes the classics; "Bauernbratwurst" sausage with onion sauce, veal Cordon Bleu with herbed French fries, and – naturally – Zurich "Geschnetzeltes" (stir-fried veal) served with rosti. (Walliser Keller 2019)

Rajdoot, an ethnic restaurant in London (UK) has promoted the notion of authentic experiential consumption and recreates its Indian cuisine in a manner that can be adapted to the taste of different target markets. The restaurant claims to offer the best flavor of India and conveyor of cultural messages. It promises stimulating sensory experiences and good service with trained staff, while also taking pride in its atmospheric appeal:

> We recreate traditional Indian dishes covering this wonderful, vibrant country – from the Himalayas to the Indian Ocean and the Arabian Sea to the Bay of Bengal. Rajdoot draws inspiration from all twenty-two regions and our food varies according to local influences characterised by exquisite aromas, glowing colours and earthy flavours. Our team of highly skilled chefs use only the finest ingredients, buying fresh spices which are painstakingly ground and mixed by hand, ensuring pure quality. Enjoy your meal at Rajdoot;
>
> The experience you will have when you visit The Rajdoot restaurant is more than just the tasteful cuisine. Our welcoming staffs have been

> trained in all aspects of food and beverage culture and they will ensure that you feel comfortable and taken care of. The Rajdoot draws inspiration from all twenty-two regions and our food varies according to local influences characterized by exquisite aromas, glowing colors and earthy flavors;
>
> Here we recreate traditional Indian dishes covering this wonderful, vibrant country – from the Himalayas to the Indian Ocean and the Arabian Sea to the Bay of Bengal. Our team of highly skilled chefs uses only the finest ingredients, buying fresh spices, which are painstakingly ground and mixed by hand, ensuring pure quality. Rajdoot Restaurant (2019)

Many review sites endorse the popularity and genuineness of its ethnic gastronomic experience. However, while the review sites share and praise objectively authentic experiences, the restaurant themes itself as a cultural tradeoff as evidenced in its promotional narratives. For instance, authentic taste and aromas, pure ingredients, and traditional way of cooking by hand are synergized with high quality of service and warm welcome. An effort is made to transmit the national flavor of India by embracing and mixing food heritage from various regions of the country.

Slow food journeys in the context of cultural immersion are also emerging as important topics of study (Miele & Murdoch 2002). In their paper focusing on slow food experience at a restaurant in Tuscany, the authors highlight three key significant atttributes: first, the 'organizational aesthetics' of restaurant life, where tacit knowledge, craft skills, and tradition are brought to the fore; second, the 'aesthetical ethics' of typical foods, which are promoted in these restaurants; third, the connectedness of the restaurant to its surrounding ecosystem and local economy. These three aspects all highlight the importance of a practical aesthetic, one that is rooted in the tacit knowledge, craft skills, creative energies, and socio-natural relationships that comprise typical cuisines (2002, p. 313). These considerations have been aligned with the ongoing trends associated with the production/demand for eatertainment experiences and negotiated 'othering' or 'traditionalisation' of ethnic cuisines.

It is further postulated that the consumption of ethnic food offers cultural (othered) eatertainment, a leisure experience implying a mix of entertainment and ethnic dining experience (Mokono 2007). Pan et al. argue that eatertainment experience at an ethnic restaurant is a mix of services offered in tangible and intangible settings. Ethnic eatertainment or demand for 'othered' food has become a contemporary culturally immersive phenomenon (Timothy 2011). For instance, Mkono (2013) examines review comments of visitors to ethnic restaurants in Africa and reports that most patrons seek an 'othered' experience to escape from the daily and the mundane. At this juncture, discussion will be incomplete without mention of the "othered" agents or the authenticating agencies who have the power to decide what cultural markers and expressions can be employed to transmit an authentic experience.

Numerous studies identify the authenticating agents and markers of ethnic cuisines. It is argued that authenticity of ethnic cuisines needs to be showcased and endorsed through legitimized cultural markers/symbols and agents. According to Robinson and Clifford (2012), the extent of 'otherness' varies with the manner the offered experience is marked with objectively authentic markers. Ethnic showcasing often gets hybridized in third spaces (Bhabha 1994). These spaces represent 'otherness' that is negotiated to support sociopolitical and economic agenda of host environments. But the common reference point remains object authenticity, authentication markers of which become acceptable if endorsed by legitimized institutions. Mokono (2007) argues that negotiated authenticity strategies are, for the most part, embraced by ethnic restaurants to illustrate both otherness and continuity while modifying food offerings and atmospheric settings to cater to market demand.

Robinson and Clifford (2012) also identify of a wide range of authentic symbols that align with the theoplacity view of authenticity. The authenticity markers include:

- Preparation, cooking style, and equipment used (authentic to medieval times);
- Oral and written description of menu items (food and beverage are labeled as authentic to medieval times);
- Sourcing and selection of ingredients (based on medieval times);
- Presentation of food platters and accoutrements on table;
- Use of other authenticating agents on perceived taste;
- Role playing and costume designs of service staff;
- Menus and dishes attempt to replicate what is perceived as medieval; and
- Packaging of agents of authenticity to deliver an experience distinctive of the medieval (2012, p. 14).

Chhabra et al. (2012) examine the manner in which managers of ethnic restaurants authenticate experiences and food to convey objectively authentic messages in online as well as physical settings. In a separate study, Chhabra et al. (2013) use a modified version of markers developed by Robinson and Clifford (2012) and present a universal authenticity scale for consumption of food heritage based on content analysis of fifty geographically dispersed Indian restaurants across the USA. From the supplier's side, the authenticating markers portrayed at the signature websites of select restaurants include: "traditional food and other related items (such as spices, utensils, clay oven, and tablecloth with traditional designs), traditional menu items, traditional terminology, Indian icons, and images of Indian chef and waiter/waitress.

At the same time, efforts to adapt to changing lifestyles (such as healthy) in food offerings are communicated. The restaurant owners also communicate theoplacity-negotiated narratives/visuals of otherness to instill meanings of object authenticity at their signature websites" (Chhabra et al. 2013, p. 15). It is noted that a third place is created for patrons to immerse themselves, which appeals to their comfort level. Consumers desire authentic food and its showcasing in an entertaining manner. Leisurely servicescape backdrops fulfill the 'othering' quest. Chhabra et al.'s study reinforces objective authenticity as a reference point to showcase 'othered' food experiences based on genuine attributes of Indian traditions. However, the results also highlight the need for negotiated offerings to adapt to market demand and set high-revenue targets. Most scrutinized studies report support for the theoplacity stance of authenticity that reveals efforts to balance the need for continuity and change as consumer markets become more dynamic and seek existentialist type of experiences in objectively or commodified settings.

In all aforementioned cases, negotiated authentication process is employed to rationalize authenticity. It can also be deduced that a complete and true 'othered' experience delicately integrates objectively authentic elements. Chhabra et al. (2013) shows that ethnic restaurants use a variety of authenticating markers such as menu items, gastronomic presentation, and ethnic seated arrangements. However, these studies were conducted during the pre-covid times. It is probable that the aftershocks of the pandemic will give the ethnic restaurants an opportunity to rebuild/reshape their images in a more sustainable manner. The authenticity position and the authentication process may take a different shape during the post-pandemic era. In due course of time, the demand for ethnic cuisines should surge when the world embraces the new normal or returns to the old normal. The hope is that it will not be 'business as usual' and that this pandemic will open doors for reformative solutions that celebrate pastness of authenticity in ethnic cuisines in a simpler manner.

Closing comments

This chapter, for the most part, highlights recent pre-pandemic trends associated with the negotiated 'othering' or objective authenticity of ethnic cuisines in diasporic settings. Extant literature opines that food knowledge and habits of migrants get transformed to a certain extent as they make efforts to adapt to the migrating environment (Kuhn, Haselmair, Pirker, & Vogl 2018, p. 9). They are also subject to the availability of authentic ingredients. Ethnic food in a foreign country can often be restricted to a non-representative selection of 'symbolic dishes' in a national repertoire and these often go through a commodification process through strategic selection of authenticating markers. According to Ngyyen (2007), the diaspora populations often substitute their food ingredients with local resources which are of similar taste and aroma and try to maintain the core elements of traditional foods.

When people have limited sources to check on authenticity, the managers and staff of ethnic restaurants hold the power to introduce innovative dishes to meet market demand; and they can transform dishes based on selected knowledge of food culture while holding and reinforcing preferred food traditions. Customers can be gullible and duped by the showcased promise of authentication as they might not be able to tell the difference between what is authentic and what is blended based on preferred taste (Mohammad & Chan 2011, p. 458).

Several markers are considered legitimate to authenticate ethnic food heritage. For instance, authentication is accepted by consumers if they are convinced that the authenticating agents are from the place of origin and have traditional knowledge in addition to the endorsement, by other patrons, posted on credible review sites. A case in point is the Rajdoot restaurant mentioned earlier in the chapter. Internet has offered a unique and challenging platform to sell and communicate authenticity and showcase convincing narratives/markers of authentication. Callon and Muniesa (2005) argue that the burden of authentication does not fall on a single agent as multiple actors are involved and use their power to create and communicate performative practices to reiterate preferred value claims.

Socio-technological compositions have added another qualifying layer. In this regard, Lugosi explains:

- Authentication in tourism relies on the creation of identities to be evaluated and the creating of objects is objectification – which is thus a representation of people, places, gestures, physical objects, sensations, texts, and images;
- Computing and technological algorithms are key to this objectification process insofar as they delineate the object by (en)coding data into indexable packages, making them distinct from other experiential objects, yet organizing them so they can be encountered and valued alongside others;
- Consumer authentication is not restricted to tourists and their suggestions on how future travelers can enhance their experiences. Study of online food communities reaffirms the credibility of the venue and provide advice on how visitors can refine such experiences through personalization (Lugosi et al. 2012);
- Cocreated performance of expertise reflects growing consumer empowerment and diffused power in valuation processes. However, consumer performance of expertise is subject to appropriation and (re)enrollment by organizations in their value-loaded (Callon and Muniesa 2005); and
- Discriminating between reviewers, based on their experience and the quality of their advice, can help authenticate their credibility. This in turn legitimatizes reviewers' ability to authenticate experiential objects in the future. Content providers may reinforce this qualification process by showing how many posts users have made and how their posts are rated by other users (2016, pp. 101–107).

In other words, it is important that both suppliers and consumers interact closely to cocreate the aestheticization of atmospherics and food heritage appraisal codes. The concept of aesthetication can be delineated into 'aesthetic of entertainment' (turning attention to atmospherics) and 'gastronomic aesthetics of food' (centering attention on quality of food consumed and food heritage) (Miele & Murdoch 2002). Based on a purview of documented literature, this chapter notes a mix of hot and cool authentication in ethnic restaurant settings. These also mirror Mohammad and Chan's four proposed dimensions of authenticity: Traditional long-term practice reflecting heritage and style persistence, nostalgia, genuineness (unadulterated), and assurance (2011, p. 464). Taste and the production process (spirit) are also deemed important for repeat patrons. Both supply and demand perspectives of authenticity and its authentication have echoed support for the negotiated notion. Authentic showcasing of distant culture showcasing is a complex task because of the fluid nature of authenticity notions and its vulnerability in the need to embrace market demand and sociopolitical/economic agendas. Authenticating alternative hedonism or eatertainment to influence different stages of the consumer purchase and decision-making process requires customized showcasing in promotional materials, particularly via online platforms.

With the unprecedented infiltration of information communication technology (ICT) in every sphere of our lives, online marketing of authenticity and its authentication forms an important area of study. Having said that, the ultimate experience happens in a physical and tangible setting adorned by intangible ethnic markers. Therefore, an alignment, delicate balance, and resonance between online content and onsite experience is crucial. In other words, the digital messages can enrich cognizance which can result in an immersive onsite experience. As an instance, slow heritage food experience can encompass all stages of the consumer decision-making process including the pre- and post-purchase phases by engaging the cognitive and affective senses of the consumer.

Another important aspect for consideration is cultural politics associated with the authenticity and authentication (traditionalization) of ethnic cuisines. Long (2018) argues that cultural politics in the context of ethnic foods in foreign environments are complex as cuisines, dishes, and ingredients belong to a heritage outside the foodways of the mainstream culture. Watson and Caldwell (2009) argue that food is a political window as food practices are appropriated by dynamics of dissonant, negotiated and asymmetrical relationships, expectations, and preferences. Here, cultural politics refer to the practical consequences of power such as geographic access or its pricing. As Long writes, cultural politics of food refers to issues of authority, that is, who holds the power to bestow meanings and select cultural expressions of food and eating; also, how is this authority used? Food is not only a commodity but it also has emotional connotations and is a carrier of identity and culture: as such, defining and representing food highlights the variety of interpretations that can be given to those meanings. Selecting interpretations then reflect issues of power, that is, who has the authority to make those decisions and why. Therefore, a crucial step to address the cultural politics issue is to recognize the complex nature of food and its multiple connotations, its interweaving with and mirroring different expressions of life and its multiple connotations (2018, pp. 317, 322).

While these intra-pandemic times have imposed social distancing measures and restricting gatherings in restaurants, a study shows that ethnic food continues to remain in demand, although the nature of demand is transforming. Ethnic grocery stores have evidenced a surge in their sales as coronavirus cases spiked across the globe. Many ethnic restaurants are offering curb-side pickup and home delivery services (Meehan 2020). Undoubtedly, most have shrunk their menus and are offering limited choices. However, they can introduce more economical and traditional choices and innovatively market authenticity of their cuisines on their websites. Many consumers dine in to socialize, dress up, and catch up with friends and relatives in addition to patronizing authentic ethnic food and atmospherics (Sannchez 2020). With the new normal, attention is shifted more towards quality of food and interest in its pastness and historicity because the social diversions and décor markers are less likely to be sought until the pandemic is controlled with a vaccine. Ethnic restaurants will continue to develop resilient techniques to remain in demand, particularly, by more delicately negotiating cultural authenticity of ethnic cuisines and its authentication process.

In summary, this chapter places emphasis on the authentication of negotiated ethnic experiences that resonate with the lifestyle of heritage tourism markets. It draws attention to the mediating platforms and the cultural politics that shape the deconstruction of authentication of ethnic cuisines in foreign environments. It is paramount to unmask the socio/political dynamics of processes that authenticate ethnic food production and consumption. A pluralistic framework should be in place for the purpose of delineation recognizing the multifaceted nature of foreign settings and different (working through interrelated and mutually entangled systems) agents of authentication. As pointed out by Cohen and Cohen, politics of authentication is not a zero-sum game, but the whole meaning making and showcasing of negotiated authenticity is a 'constitutive performative process' (2012, p. 128). The pandemic presents unique set of challenges because of mobility and social distancing restrictions. In the next chapter (Chapter 11), I scrutinize another unique facet of tangible heritage (souvenirs) and offer insights on the souveniring process and its negotiated authentication to resonate with the changing times.

References

Andersson, T., & Mossberg, L. (2004). The dining experience: Do restaurants satisfy customer needs? *Food Service Technology*, *4*, 171–177.

Barbas, S. (2003). I'll take chop suey: Restaurants as agents of culinary and cultural heritage. *Journal of Popular Culture*, *36*(4), 669–686.

Bhabha, H. (1994). 17 Frontlines/Borderposts. *Displacements: Cultural identities in question*, *15*, 269.

Callon, M., & Muniesa, F. (2005). Economic markets as calculative collective devices. *Organization Studies*, *26*(8), 1229–1250.

Chaney, S., & Ryan, C. (2012). Analyzing the evolution of Singapore's World Gourmet Summit: An example of gastronomic tourism. *International Journal of Hospitality Management*, *31*(2), 309–318.

Chhabra, D., Lee, W., Zhao, S., & Scott, K. (2013). Marketing of ethnic food experiences: Authentication analysis of Indian cuisine abroad. *Journal of Heritage Tourism*, *8*(2–3), 145–157.

Chhabra, D., Zhao, S., Lee, W., & Okamoto, N. (2012). Negotiated self-authenticated experience and homeland travel loyalty: Implications for relationship marketing. *Anatolia*, *23*(3), 429–436.

Cohen, E. (2007). Authenticity in tourism studies: Apres Le lutte. *Tourism Recreation Research*, *32*(2), 75–82.

Cohen, E., & Cohen, S. A. (2012). Authentication: Hot and cool. *Annals of Tourism Research*, *39*(3), 1295–1314.

Game, A. (1991). *Undoing the social: Towards a deconstructive sociology*. Milton Keyes: Open University Press.

Guerrero, L., Guàrdia, M. D., Xicola, J., Verbeke, W., Vanhonacker, F., Zakowska-Biemans, S., & Scalvedi, M. L. (2009). Consumer-driven definition of traditional food products and innovation in traditional foods. A qualitative cross-cultural study. *Appetite*, *52*(2), 345–354.

Henderson, J. C. (2012). Conserving heritage in South East Asia: Cases from Malaysia, Singapore and the Philippines. *Tourism Recreation Research*, *37*(1), 47–55.

Henderson, L. (2007). "Ebony Jr!" and "Soul Food": The Construction of Middle-Class African American Identity through the use of traditional southern foodways. *Melus*, *32*(4), 81–97.

Horng, J. S., & Tsai, C. T. (2012). Culinary tourism strategic development: An Asia-Pacific perspective. *International Journal of Tourism Research*, *14*(1), 40–55.

Jang, S., Liu, Y., & Namkung, Y. (2011). Effects of authentic atmospherics in ethnic restaurants: Investigating Chinese restaurants. *International Journal of Contemporary Hospitality Management*, *23*(5), 662–680.

Kuhn, E., Haselmair, R., Pirker, H., & Vogl, C. R. (2018). The role of ethnic tourism in the food knowledge tradition of Tyrolean migrants in Treze Tílias, SC, Brazil. *Journal of Ethnobiology and Ethnomedicine*, *14*(1), 26.

Lai, I., Lu, D., & Liu, Y. (2019). Experience economy in ethnic cuisine: A case of Chengdu cuisine. *British Food Journal*, doi: 10.1108/BFJ-08-2018-0517.

Liao, S., & Ma, Y. (2009). Conceptualizing consumer need for product authenticity. *International Journal of Business and Information*, *4*(4).

Liu, H., Li, H., DiPietro, R. B., & Levitt, J. A. (2018). The role of authenticity in mainstream ethnic restaurants: Evidence from an independent full-service Italian restaurant. *International Journal of Contemporary Hospitality Management*, *30*(2), 1035–1053.

Long, L. (1998). Culinary tourism: A folkloristic perspective on eating and otherness. *Southern Folklore Quarterly*, *55*(3), 181–204.

Long, L. (2018). Cultural politics in culinary tourism with ethnic foods. *Revista de Administração de Empresas*, *58*(3), 316–324.

Lu, S., & Fine, G. A. (1995). The presentation of ethnic authenticity: Chinese food as a social accomplishment. *The Sociological Quarterly*, *36*(3), 535–553.

Lugosi, P., Janta, H., & Watson, P. (2012). Investigative management and consumer research on the internet. International Journal of Contemporary Hospitality Management.

McCoy, L. (2012). Food Heritage Planning Proposals: Planning For Food Heritage Celebrations In Central Virginia. Retrieved from http://www. virginia. edu.

Meehan, P. (2020). Eating in the time of coronavirus. Retrieved in April, 2020, from https://www.latimes.com/food/story/2020-03-12/coronavirus-eating-restaurants-dont-be-racist.

Miele, M., & Murdoch, J. (2002). The practical aesthetics of traditional cuisines: Slow food in tuscany. *Sociologia Ruralis*, *42*(4), 312–328.

Mintz, S. W. (1996). *Tasting food, tasting freedom: Excursions into eating, culture, and the past*. Beacon Press.

Mkono, M. (2013). Using net-based ethnography (ethnography) to understand the staging and marketing of 'authentic African' dining experiences to tourists at Victoria Falls. *Journal of Hospitality and Tourism Research*, *37*, 184–198.

Mohammad, T., & Chan, J. K. L. (2011, July). Authenticity representation of Malay Kelantan ethnic cuisine. In *The 2nd International Research Symposium in Service Management, Yogyakarta, Indonesia*.

Mokono, M. (2007). The othering of food in touristic entertainment: A netnography. *Tourist Studies*, *11*(3), 253–270.

Morrison, A., Lehto, X., & Day, J. (2018). *The tourism system* (8th ed.). Dubuque (IA): Kendall Hunt Publishing Company.

Naccarato, P., & LeBesco, K. (2013). *Culinary capital*. Bloomsbury Publishing.

Ngyyen, M. (2007). Community dynamics and functional stability: A recipe for cultural adaptation and continuity. *Economic Botany*, *61*(4), 337–346.

Okumus, B., Okumus, F., & McKercher, B. (2007). Incorporating local and international cuisines in the marketing of tourism destinations: The cases of Hong Kong and Turkey. *Tourism Management*, *28*(1), 253–261.

Omar, M., Mohd Adzahan, N., Mohd Ghazali, H., Karim, R., Abdul Halim, N. M., & Ab Karim, S. (2011). Sustaining traditional food: Consumers' perceptions on physical characteristics of Keropok Lekor or fish snack. *International Food Research Journal*, *18*(1), 117–124.

Pratt, J. (2007). Food values: The local and the authentic. *Critique of Anthropology*, *35*, 35–37.

Rajdoot Restaurant (2019). Retrieved on October 2, 2019, from https://www.therajdootindian.com/.

Ramli, A., Zahari, M., Ishak, N., & Sharif, M. (2014). Food heritage and nation food identity formation. *Hospitality and Tourism: Synergizing Creativity and Innovation in Research*, *407*(2013).

Robinson, R., & Clifford, C. (2012). Authenticity and festival foodscape experiences. *Annals of Tourism Research*, *39*(2), 571–600.

Sannchez, B. (2020). Coronavirus and restaurants: what diners miss the most during quarantine. Retrieved in June, 2020, from https://www.thedailymeal.com/what-people-miss-about-restaurants-coronavirus-quarantine.

Shariff, N., Mokhtar, K., & Zakaria, Z. (2008). Issues in the preservation of traditional cuisines: A case study in Northern Malaysia. *The International Journal of the Humanities*, *6*(6).

Sims, R. (2009). E-service quality: A model of virtual service quality dimensions. *Journal of Sustainable Tourism*, *17*(3), 321–336.

Song, H., Phan, B., & Kim, J. (2019). The congruity between social factors and theme of ethnic restaurants: Its impact on customer's perceived authenticity and behavioral intentions. *Journal of Hospitality and Tourism Management*, *40*, 11–20.

Stiles, K., Altıok, Ö, & Bell, M. (2011). The ghosts of taste: Food and the cultural politics of authenticity. *Agriculture and Human Values*, *28*(2), 225–236.

Stock, M., & Schmiz, A. (2019). Catering authenticities: Ethnic food entrepreneurs as agents in Berlin's gentrification. City, Culture and Society. Retrieved from https://doi.org/10.016/j.ccs.2019.05.001.

Timothy, D. (2011). *Cultural and heritage tourism*. Bristol: ChannelView Publications.

Tsai, C., & Lu, P. (2012). Authentic dining experience in ethnic restaurants. *International Journal of Hospitality Management*, *31*, 304–306.

Wahid, N. A., Mohamed, B., & Sirat, M. (2009, July). Heritage food tourism: Bahulu attracts. In *Proceedings of 2nd National Symposium on Tourism Research: Theories and Applications* (pp. 203–209).

Walliser Keller (2019). Retrieved on October 4, 2019, from https://walliser-keller.ch/en/.

Watson, J., & Caldwell, M. (Eds.). (2009). *The cultural politics of food and eating: A reader*. Oxford, UK: Blackwell.

Yang, L., Wall, G., & Smith, S. (2008). Ethnic tourism development: Chinese government perspectives. *Annals of Tourism Research*, *35*(3), 751–771.

11 Souveniring of heritage souvenirs

This chapter examines the manner in which authenticity of souvenirs and handicrafts is negotiated and offers a discursive view of the process of souveniring (authenticating). Furthermore, it offers critical insights on how these popular categories of heritage merchandize undergo digital authentication. Insights are also shared on how several heritage agencies are innovatively adapting to the COVID-19 challenges and offering opportunities for digital cultural engagement/immersion to their audience and other extended markets.

Shopping has been one of the most popularly sought out activities (Timothy 2011) in heritage tourism as it offers a complete heritage experience (Haldrup 2017; Cohen 2000; Hitchcock & Teague 2000). This activity has not escaped the wrath of the pandemic as the heritage tourism industry pauses. Many souvenir shops in museums and heritage villages/sites have closed their doors or restricted physical movement inside their premises. However, due to the resilient nature of heritage tourism, creative alternatives can be evidenced in the horizon that echo and resonate with the utility and emotions of covid times. As evidenced from Marshall's narrative, creative and culturally designed mass coverings have become immensely popular in terms of cultural and artistic connections. A dialogical view on authenticity and authentication from the pre-covid era is extended to the world of souvenirs in this chapter. Toward the end, insights are shared on how several heritage agencies are innovatively adapting to the COVID-19 challenges and offering opportunities for digital cultural engagement/immersion to their audience and other extended markets.

It cannot be denied that heritage tourism merchandize is popularly sought in destinations at "shopping malls, festivals and events, museums and historic houses, rural and urban villages, hotels and resorts, conference centers, airport shops, bus station shops, cruise ships, railway stations, the Internet, and cross border venues" (Chhabra 2010, p. 164). In fact, shopping of heritage merchandize is symptomatic of a cultural medium through which cultural values of a destination or community can be captured (Halewood & Hannam 2001). It is influenced by several motives: status, nostalgia, less expensive, quest for authenticity, utilitarian needs, altruism, keeps busy, gift for others, and cultural obligations (Haldrup 2107; Timothy 2011).

Majority of the heritage tourists particularly enjoy buying souvenirs during their travel (Silberberg 1995; Timothy 2011). Souvenirs capture uniqueness of a destination and are a popular shopping attraction. Return from travel with a souvenir serves as a "reminder of special moments or events" (Swanson 2004, p. 363) and "few tourists come home from vacation without something to show ....as proof that they really did make the journey" (Graburn 1987, p. 395).

Souvenirs perform a central role in enhancing travel experience and are the very essence of cultural showcasing and therefore the most commonly purchased category of heritage merchandize (Gordon 1986b; Haldrup 2017). The souvenir phenomena can be best described by the sacredness theory that refers to the sacred state of individuals who travel away from their everyday life to experience something novel, extraordinary, and exotic (Gordon 1986a). Because their visit is temporary, they bring home, a tangible piece of the extraordinary, to remind themselves of their visit and relive their experience (Fu, Liu, Wang, & Chao 2018; Haldrup 2017). This chapter examines the manner in which authenticity of souvenirs and handicrafts is negotiated and offers a discursive view on the process of souveniring (authenticating). Furthermore, it offers critical insights on how these popular categories of heritage merchandize undergo digital authentication. The core focus of this chapter is on:

- Negotiated authenticity of souvenirs
- The souveniring process
- Digital authentication of souvenirs

Souvenirs can be described as material mementos, popular keepsakes, and touchstones of memory that recite and revive stories, of the distant era, in a way that they can be relived or reexperienced (Haldrup 2017; Morgan & Pritchard 2005). They range from primitive handicrafts to mass-manufactured items. The commonly used classification of souvenirs is:

- Pictorial – examples include postcards, books, and photographs;
- Piece-of-rock – examples include rocks, shells, wood, fossils, bones, and pinecones;
- Symbolic shorthand – can be replicas of famous attractions and manufactured items with images of the visited place;
- Markers – utility items are marked with words and logos such as coffee mugs, coasters, key chains, and shot glasses;
- Local products – are local merchandize such as food, drinks, cooking utensils, clothing, and handicrafts; and
- Non-local products – such as non-local handicrafts and craftwork category. These are imported from ancestral home countries of diaspora populations (Chhabra 2010; Gordon 1986b).

Handicrafts and ancient art pieces are important category of souvenirs exemplifying the bygone eras as they symbolize sociocultural assets of visited communities. In fact, handicrafts and art pieces form the most generic category of souvenirs (Timothy 2005). Handicrafts are made by hand and are enshrined with tradition designs and raw materials. Popular examples include textiles, pottery, baskets, figurines, wood carvings, and jewelry. Art in the Americas has been broadly classified into four categories based on its meaning, utility, artist perceptions, and the relationship the artist has with the consumer (Feest 1992):

- Tribal art – this form serves functional utility for the tribal groups. The core purpose of its production is its usefulness.
- Ethnic art – this kind of artwork is created for use by other ethnic groups. It has become a source of revenue and also a symbol of the artisan's ethnic identity.
- Pan-Indian art – this form is the outcome of demand external to the local region. It is made to meet that demand and is modified to match consumer needs.
- Indian mainstream art – this work is created by artists who just happen to be Indians. Each artist has his/her own expressions and the theme of their work may be based in part at least on their ethnic heritage (Timothy 2005, p. 107).

Handicrafts have, in fact, become the largest source of income in the developing world after agriculture. Historically, they were produced to appease the utilitarian and ceremonial needs of a host community (Popelka & Littrell 1991). Economic necessity compelled craft producers to seek alternate venues and target heritage tourism markets (Cappucci 2016; Kord, Heidari, & Rigi 2016). Multiple factors, both internal and external, changed the form and functionality of traditional handicrafts and arts thereby opening the path for their commodification (Cohen 2000). Airport art or tourist art is the outcome of this commodification process (Graburn 1976, 1982).

Documented literature presents four strands of research on souvenirs: producer and vendor strategies, consumer preferences and behavior (Altintzoglou, Heide, & Borch 2016), perceived authenticity of souvenirs, and digitalization of souvenirs:

- In the first strand, researchers have examined marketing/branding strategies of producers and retailers (Chhabra 2005, 2010, 2015; Luscombe, Walby, & Piché 2018; Thomsen 2018).
- The second strand highlights experiential consumption of souvenirs, their memories, and emotive appeal/connection (Fu et al. 2018; Haldrup 2017). Focus is also on the meanings tourists attach to souvenirs and the manner in which they become coinhabitants at home (Fu et al. 2018; Haldrup 2017; Littrell 1990). According to Morgan and Pritchard, "souvenirs may acquire new meanings and significances when they move through time and spaces of the people they live with, carrying with them a complex and changing social life of their own" (2005, p. 45).

- The third strand offers a dialogical view on debates surrounding souvenir authenticity and its perceptions (Cohen 2000; Chhabra 2005, 2010 Hitchcock & Teague 2000; Littrell, Andersen, & Brown 1993).
- The last strand is increasingly gathering prominence as heritage agencies embrace digitalization to pursue innovative ways to produce, engage visitors, and offer interactive experiences with souvenirs.

According to Esperanza (2008), retailers as middlemen (acting as mediating agents between the producers and the consumers) make significant interventions in the production and sale of souvenirs in a variety of ways such as: embracing innovative methods of production, suggesting different materials to artisans, preferring certain styles, and being selective of objects for external distribution. Cohen (2000) posits that local crafts can change or become commercialized because of social, political, and economic mediations such as war and by embracing or integrating styles from other cultures etc. Cohen does not agree that external demand for the commercialization of arts/crafts compromises the vitality and authenticity of past traditions, thereby rendering the commodified products unusable to the locals and the producers. According to him, such processes "are only another stage in their historical process of change" (Cohen 2000, p. 4). Four dichotomous facets explain the complex nature of these dynamic processes:

1 Perpetuation vs. Innovation – These relate to the extent to which the art or craft changes that it loses its original elements. That is, degree to which reproduction is tended or new elements change the nature of the original structure;
2 Orthogenesis vs. Heterogenesis – The point of attention here is the nature of the 'stylistic Innovations'. Either inherent traditional styles continue to be used by artisans or new radical style elements are introduced which are not connected with the general cultural rituals of the artisan;
3 Internal vs. External audience –The first type of audience holds similar cultural values with the artisans, while the external audience is not connected with the local cultural values; and
4 Spontaneous vs. Sponsored production – The first type refers to self-initiative toward commercialization, whereas the sponsored production is driven by an external agency such as a government or non-government organization or a private entrepreneur (Chhabra 2010, p. 168; Cohen 2000).

Ethic art production has always been in a state of negotiation and transformation due to external influences, factors, and mediating human agencies. Esperanza (2008) refers to the middlemen as cultural custodians and opines that these mediators not only channel the passage of these commodities, but also authenticate their cultural values and meanings in multiple ways. The author uses the phrase "outsourcing of otherness" to describe how local artisans in Bali are contracted to continue the work of imagining 'otherness'; to continue the simulacrum of subaltern culture imagined by those who hold economic and political influence (Esperanza 2008, p. 74). External buyers are gullible to the cultural and historical narratives of producers who employ selected cultural/traditional materials and markers to produce and authenticate souvenirs (Paraskevaidis & Andriotis 2015). Affluent heritage knowledge is valuable to enrich them with a touch of a mythical or folk tale so that they can be prized as more objectively authentic to augment their market and sale value (Ciftki & Walker 2017).

Negotiation of authenticity

Undeniably, objective authenticity remains the point of reference, and the most important authenticating point remains 'the place of origin' and 'made by local hands' (Asplet & Cooper 2000, p. 308). However, in heritage tourism markets, authenticity is a fluid phenomenon. Cohen writes that authenticity is socially negotiable because "authenticity in the objects we buy and study is seen as a sign of our own alienation. Or as a means of preserving their own historicity" (1993, p. 142). Therefore, the authenticity of historical markers and heritage objects continues to be authenticated as it goes through a process of revision, recast, and reinvigoration. Hollinshead (1996) adds that history and culture have always evolved and therefore authenticity will continue to be a fluid and negotiated phenomenon. In support for the negotiated stance of authenticity, one can also argue that authenticity is not a tangible asset but a judgment or value conferred on a product (Moscardo & Pearce 1999; Xie & Wall 2002). The underpinnings of this argument are that tourists are active authenticating agents and consumers of authenticity.

From an opposite perspective, however, it has also been posited that heritage suppliers are the authenticators. Most of the research on suppliers/authenticating agents of heritage merchandize has centered on the manner they theme and package authenticity for sale. Several factors guide the continuous production of authentic products such as persistent demand, availability of traditional materials, existence of workforce, knowledge, and ongoing utility of the product in the daily life of the host communities (Ciftci & Walker 2017; Graburn 1982; Haro Zea, Haro-Zea, Roblero-Mazariegos, & Sánchez Sánchez 2018; Maghsoodi & Nadalian 2018; Schilar & Keskitalo 2018). Producers employ innovative techniques to uphold the authenticity of their crafts to tourists. Reville and Dodd (2003) report that producers often deliberately make an object exclusive and hard to find to confer an authentic status. Some use the market skimming pricing strategy to authenticate its intrinsic qualities. Also, there is a contemporary trend toward the production of new kinds of functional objects, which are modified based on personal significance and lifestyles of the targeted audience

(Cohen 1993; Fu et al. 2018; Kord et al. 2016; Maghsoodi & Nadalian 2018; Haro Zea et al. 2018).

Aligned with the above, contemporary handicrafts should coexist innovatively so that they continue to be rooted in local traditions. Authenticating criteria for handicrafts include: rootedness in local culture, historic, having an identity, connected to their original location, and to their artist (Fu et al. 2018). Fu et al. write that handicrafts are not accountable to a monolithic closed concept. They are spontaneous and adaptive and hold extraordinary power to express ideas and transmit emotions, although they emerge from limited resources; they embrace fresh creativity and evolve with time. For instance: "in the State of Chiapas, a place recognized for its cultural tourism, handicrafts are not a complimentary element, but an indispensable element to achieve competitive advantage.... and in the context of Chiapanecan handicrafts, the nature of the handicrafts allows artisans to have a sustainable competitive advantage due to their being handed down from generation-to-generation, as well as creativity being present ... ... through ... cultures, traditions, emotions, and the feelings of their creators" (pp. 335, 341).

Recognizing a negotiated authenticity standpoint for handicrafts from Eastern Turkey, it is argued that locality is important not just as a physical setting where sales happen but can be optimized as an innovative agent (Ciftci & Walker 2017)). However, it is imperative not to compromise the custom of coexisting with our natural and cultural resources to commodify locality and local production (Birkeland 2002). Ciftci and Walker lend credence to a negotiated strategy that "invigorates traditional crafts production, marketability, and promotion" (2017, p. 3002). Chhabra also reports emphasis on negotiated authenticity in her scrutiny of Kashmiri shawls (stolls) from North India:

> As far as authenticity is concerned, the Kashmiri vendor today has a mixed story to tell. Even though genuine Kashmir shawls are still available and some of them still maintain the traditional values, paucity of both labor and genuine material, development of education, and easier accessibility of wider markets have encouraged many of them to sell spurious in the name of genuine products to the ignorant and unsuspecting visitors.

Different versions of authenticity are noted between the vendors and the artisans, the local community, and the government. The locals sell modified versions as heritage items to lure tourists to visit Kashmir (Chhabra 2010, p. 177).

The souveniring process

Proceeding forward, this chapter angles on the souveniring route to offer insights on how a preferred version of authenticity is bestowed on a souvenir. As pointed out earlier, the special value of a souvenir lies upon return home, where it rests as a 'touchstone' of memory and serves to remind of 'extraordinary' travel moments. It is important to note that souvenirs are subject to dual and sometimes conflicting authentication paths. One route is orchestrated by the heritage agencies/institutions whereas the other path is authenticated by the tourists themselves through their inner selfing. Documented literature shows that the act of souveniring is a form of negotiated authentication that takes place between the place, the object and the consumer and within the consumer. The second path can be described as existentialist authentication, which strives to elevate 'a sense of being' (Lausa 2007) "locked into the memory of the collection and the collector's experience of the site-date of production, name of the producer and monetary value are to some extent immaterial" (Hume 2013 cited in Cave & Buda 2018). Emotional values associated with the place are infused in the souvenirs that heighten authentic experiential consumption (Fu et al. 2018; Tsai & Wang 2017). Undeniably, souvenirs are emblems of negotiated authenticity and they are popularly souvenired or authenticated in a sociocultural and emotional manner with the use of local materials and local producers. Both, the personal experience of a tourist and the host culture of the visited place become meaningful through tangibilization and sanctification of elevated objectively authentic perceptions (Fu et al. 2018). Therefore, souvenirs are realistically imbued with negotiated authenticity because they are the outcome of social-construction of emotions.

On a positive note, negotiated authenticity can dilute constructivist manifestations, with the latter referring to stereotyped fakeries and pseudo productions by non-local producers who seek to market a destination in a preferred way. In negotiated endeavors, an artisan can mix and stabilize change with traditional continuity. As an example, high-priced historically authenticated souvenirs are displayed next to low-priced mass-produced fake objects (Cave & Buda 2018). In this manner, objectively authenticated objects are showcased next to pseudo items, which offer a comparative expression. Further to that, factors such as glocalization and globalization can shape the negotiated authenticity of souvenirs and their souveniring. Particularly, glocalization is worthy of scrutiny as it itself is an ongoing negotiated phenomenon that simultaneously embraces multiple sociocultural processes and performative aspects of authentication. Cave and Buda write that:

- The concept of glocalisation recognizes the pitfalls of oversimplifying complex social, cultural, spatial, and economic processes involved in souveniring. On the global-local continuum, souvenirs can relate to both the universality and contextuality of tourism markets;
- The souvenir form is 'agreed' communally as a cultural expression of a destination or experience and conforms to 'traditional style', and is authorized and sustained by the community. These may be produced in traditional and modern materials.... or in the form of

minitiarized originals. New versions 'protect the original' and are purposely produced for sharing;

- Commodification of materials happens through a creative corporeal and object-centered process of tourist encounters, wherein tourists are strolling through different spaces, sensing, bargaining, and .... these negotiated processes are as much socio-cultural as they are emotional, whereby feeling of the corporeal connects the material not only to the social, but also to the emotional and affective implications of souveniring;
- Recent theories turn away from the dematerialized tourist landscape and see spaces (sights, places, markets, and so forth) as performed consequences of human-material interactions. Travelers interact with the material environment and ...... collect and carry souvenirs home to authenticate their experience;
- The production of souvenirs can also be used by cultural agents to resist, respond, and interpret global influences at local levels, but by enacting the processes of glocalisation, they actively preserve ...... craft traditions, ... community relationships and economies (2018, pp. 707–716).

Authenticating agents and the direction of authenticity flow also constitute a notable field of inquiry. Chhabra (2005) examines the direction of the authenticity flow of Scottish merchandize between the producers, retailers, and the consumers. Insights are offered on who authenticates and how the flow is authenticated. It is noted that authentication is bartered between different role players. Using a dark tourism setting, Table 11.1 illustrates the manner in which souvenirs are the outcome of bargained authentication with a purpose to enrich experience. It can also be contended that souvenirs are commodities of emotional experiences and hold multiple appeal values (such as spiritual/intrinsic, use, exchange, and sign) (Paraskevaidis & Andriotis 2015). In this regard, three research streams can be noted in documented literature:

1 Souvenirs as holders of meanings that embody object-person-place relationship and function as props, evidence, memory, and substitute.
2 Souvenirs as tradable commodities that can be examined from multiple perspectives such as the producers, distributors/retailers, and consumers.
3 Souvenirs more generally, as the commodified manifestations of material culture, that can express the importance and value of the souvenir economy to tourism as whole. The study of souvenirs and souveniring spans multiple disciplines together with research into shopping, retailing handicrafts, authenticity, and consumption and may be examined through the lens of aesthetics and economies (Cave & Buda 2018, p. 716).

Table 11.1 The Hague and the Atlantic Wall

The exhibition was conceived as a stand-alone setting to which technology was added as an enrichment. The Atlantic Wall was a set of defensive lines and placements that were built by the German forces during World War II along the coast of the Atlantic Ocean and North Sea, from the French/Spanish border to Norway. The presence of these fortifications affected several countries and cities and their residents.
The exhibition specifically focused on the Atlantic Wall in The Hague. The exhibition dealt with the impact of the construction of the wall on the city and its inhabitants. Subjects such as evacuation and daily life in an occupied city that had become a military fortress were important in the exhibition's storyline. The story was told by means of museum objects, documents, maps, models, photographs, and videos all displayed in an evocative environment. The exhibition spaces were arranged to correspond to a physical map of The Hague, and showcases were marked as locations within the city.
Ten interactive display cases were spread across a single open-plan space, along with an introductory video projection. To enrich the visit experience, a set of personal stories were prepared by the curators and recorded using actors in order to provide affective content to complement the factual, more traditional textual descriptions. Playback of personal stories was triggered by smart replicas, reproductions of original exhibition objects augmented with digital technology (in this case, Near Field Communication (NFC) tags). Each replica represented a single perspective on the story of the Atlantic Wall that of a Civilian displaced from their home, a Civil Servant (Official) unwilling to be the instrument of the implementation of the occupiers' plan, or an occupying German soldier who felt himself to be a protector of the Dutch population against potential invaders.
The exhibition had textual information in two languages, Dutch and English, and therefore two sets of three smart replicas were created. Each replica represented one of three perspectives in one of the two languages. When entering the exhibition, the visitor could choose one of these replicas and carry it around the exhibition to interact with the display cases. If, for example, the visitor chose the dictionary for their visit, it would be the German perspective in English. Each replica contains an NFC tag that allowed the physical item and its user to interact with the display cases. Next to each display case was a pulsing orange circle under which was placed an NFC reader. When a replica was placed on the circle, the color changed to a solid green, indicating that the case has been activated and audio played through the earpiece, while images were projected on the case front glass. The use of replicas was monitored and information logged: when replica was placed on an NFC reader, the showcase number, the perspective represented by the replica, and a timestamp were logged; similar information was recorded when the replica was moved thus enabling the system to fully track the visit.
The visitor was allowed to freely move around the exhibition visiting cases in any order they wished. When they decided to end their visit, they could go to the souvenir printing station. Here, the replica closed the visit session, triggering the system to process the log data and generate a data souvenir: a personalized card as evidence of visit. Besides, the souvenir station was an interactive tabletop displaying the online map to invite visitors to explore the stories contributed by other visitors. In this way, online visitor contributions were part of the exhibition itself. It was also expected that, by seeing other visitors' content, visitors were encouraged to consider contributing. The possibility of encouraging visitors to contribute was considered at the design phase.
The personalized card – the data souvenir, a postcard printed at the end of the visit, contained all necessary information to go online, revisit the experience, get further material available at the exhibition, and contribute content in the form of personal and family memories.

Table 11.1 The Hague and the Atlantic Wall (*Continued*)

As with other cases, at the souvenir station, the visitor placed their replica on the pulsing orange circle. A video then played with instructions and the postcard-like data souvenir could be printed. The front of the postcard was personalized with the language and perspective of the visit, along with the three cases where the visitor spent the most time. This information was extracted from the log recorded during the visit. The three most seen cases were printed as stamps as reminder passes that needed to be stamped to permit re-entry. The card included a unique passcode and prompted the user to visit the post-visit website. The back of the postcard showed a map of the city along with reference to the corresponding locations in the city. In this way, relationship was established between the exhibition and the city outside of the museum walls. This connection between the experience of 'The Hague during the war' and 'The Hague today' was reinforced by a large interactive map positioned next to the personalized card printing station.
The postcard had a unique code printed on the top right corner. Using this code and the website address provided on the data souvenir, the visitor could elect to go online to continue the interaction with the exhibition and to add their own content to this map. The online post-visit experience was designed around an interactive map of the city of The Hague; the place in the exhibition was marked on the map, and colored to show whether the place had been visited at the exhibition or whether, it offered new content to the visitor. Standard map place markers were used to show visitors' contributions; clicking on the place marker would display the content.

Source: Petrelli et al. (2017, pp. 282–286)

Clearly, souvenirs perform numerous functions. They can be mementos of a memorable experience while also serving as cultural expressions and forging emotional connections with a place and/or a community. Furthermore, they are marketable commodities as they hold economic value. The negotiated context of dark souvenirs can be seen in the example of dark tourism souvenir from Aotearoa (New Zealand), where souvenirs were made during World War II to give brooches to expeditionary American forces stationed in the country and this practice continued after the war. Soldiers in active duty also made items, for example, sweetheart brooches fashioned from badges .... These might be considered 'dark souvenirs' because they were made by soldiers, and converted into exquisite symbolic pieces exemplifying remembrance, of soldiers who survived the war and those who succumbed to the war.

Another example are souvenirs from Palestine/Israel that became an exemplified medium through historical narratives associated with sociopolitical and economic chaos are conveyed and bargained. Cave and Buda share that "Israeli and Palestinian manufacturers and retailers of souvenirs assert their identities, their claims in the region, and tell their stories. In the West Bank in Palestine, 'icons' of the ongoing Israeli-Palestinian conflict, such as the separation wall/security fences, turnstiles at checkpoints, and even refugee camps are represented in the souvenirs produced and sold to tourists.

An array of emotions such as fear, fun, and excitement along with sensory engagements of the gaze at the existing separation wall/security fence, touch of the cold turnstiles at checkpoints, and the smell of olive trees, become part of souveniring experiences in these dark tourism places. Indeed, miniaturized souvenirs act as emotional enhancements. Third example are the murder sites in USA, where, for instance, "notions of kitsch or teddy-bearification' of the 9/11 terrorist action has been accused of trivializing and politicizing the event and creating a spectacle and subjectivities.... specific victim groups may portray animosity towards tourist activity" (2018, pp. 717–719).

Souveniring is an act of acquiring and authenticating souvenirs to personify an experience. As an instance, Cave and Buda present different contexts of souveniring in the context of dark tourism, based on a review of documented literature. A variety of places exist which offer experience of "on-going socio-political chaos; post-disaster sites; recent and ancient heritage; battlefields and submerged and land-based archaeological sites; post-war cemeteries; sites of infamous murders; and sites of staged horror such as Dracula castle" (2018, p. 717).

Heritage institutions such as the museums often shape the souveniring process and authenticate selected narratives/souvenirs that frame peace and promote prevailing political ideologies.

Souvenirs, as authentic exemplars of memory, negotiate tourism experiences by recreating emotional and sensory connections during and post-visits (Morgan & Pritchard 2005). Next, worthy of scrutiny is digital authentication of souveniring, that is, the manner in which souvenirs are authenticated on digital dais.

The term 'data souvenir' is used by Petrelli, Marshall, O'brien, McEntaggart, and Gwilt (2017) to tangibilize visitors experience at a museum. The process is actively constructed from data recorded throughout the visit and therefore it captures and showcases the actual experience. The souvenir is a keepsake (reminder) of the visit and opens the path for further online engagement, for instance, by offering visitors an opportunity to verbalize their feelings and sense of connection with it. This initiative is focused on engaging the visitors during all three stages of engagement: prior to the museum visit, during the visit, and post-visit. Because the experience is not confined within the site parameters, bonding can happen at any or across all three stages: "while pre-visit is focused on providing information, e.g. about both the collection and practical issues around the visit, the post-visit aims at establishing a long-lasting relationship that can foster additional visits and word-of-mouth promotion" (Petrelli et al. 2017, p. 281). The authors note that the purpose of nourishing a longstanding connection with visitors via digital media is, for the most part, detached from the tangible visit experience of the exhibition or an historical place. Onsite interaction and ubiquitous computing can shrink this disparity "currently existing between the exhibition floor and the online services and it can be possible to design visitor experiences that take both aspects into account. The digital and material can become components of a holistic visitor experience that crosses the digital-material boundary. The challenge is weaving the digital and the material to create seamless immersive and novel visitors' experience" (Petrelli et al. 2017, p. 282). Using a popular museum exhibit, Table 11.1 offers an illustration on how souveniring is authenticated by combining the material and digital platforms.

The aforementioned is an interesting example of how souveniring process is authenticated with the use of multiple tools to facilitate interactivity and learning. Furthermore, one can argue that the concept of data souvenir enriches reflection and memory, thereby offering an opportunity to prolong engagement beyond the museum walls. Presenting the old Hague in the context of the present Hague situates the visitor realistically in the contemporary context, while at the same time offering opportunities for time travel in an active/engaging manner. The exhibition is authenticated meticulously to avoid information and visual overload; it offers different layers of information and degrees of interactivity as desired by the visitors.

Clearly, the authentication process of the data souvenir can be guided and managed in a strategic manner. In this example, first, it is ensured that the selected souvenir resonates with the exhibition theme and offers layered connectivity through "representation of some aspects of their visit, the connection between the displays within the exhibition and the city of The Hague and access to an online post-visit experience where the visitors could find curated content as well as visitor-generated content, such as personal and family members" (Petrelli et al. 2017, p. 287).

The personalized data souvenir is automatically created and acts as a tangible reminder of a museum visit and also serves as a conduit between the site and subsequent post-online experiences. Furthering this idea of interactively engaging visitors with museums, Flynn and Flynn (2012) suggest the concept of prosumers and argue that digital innovation offers opportunities for visitors to perform the role of prosumers in the souveniring process where co-creation becomes an important and enjoyable part of the experience. Examples in this category include 3D printing technologies and open access digital fabrication.

Closing comments

Two striking aspects, outlined in this chapter, convey the demand for authentic experiential consumption and usefulness of innovative digital authentication techniques. Heritage agencies are exploring digital creativity to accentuate immersive experiential settings and optimize existentially authentic experiences. Also, an important takeaway point is the mutual authentication process through which souvenirs tangibilize visitor experience at the site and at home with their charismatic presence: "souvenirs with perceived authenticity can help tourists remember their sacred experience in the context of the host culture, the place and the time" (Fu et al. 2018, p. 356).

Co-creation becomes an important aspect in the mutual authentication process and it can be recommended that the providers "consider experiential strategies and activities that are likely to increase souvenir authenticity and the perceived value of tourists related to souvenir purchases. These activity offerings should enhance authenticity and promote communication between visitors and the local community culture. Visitors could misunderstand souvenir authenticity through not understanding the local culture. Engaging in experiential consumption could help break down these barriers" (Fu et al. 2018, p. 367).

If destinations integrate online and on-site tourist communities, they can influence souvenir purchase behavior, especially by facilitating engagement through experiential consumption. Online shopping options can be offered to deliver souvenirs used to facilitate experiential consumption (Fu et al. 2018). While deliberating on the association between cultural heritage and information communication technology (ICT), Chianese, Piccialli and Valente write that:

> the relationship between cultural heritage domain and new technologies has always been complex, dialectical and often inspired by the human desire to induce these spaces not created for that purpose, to pursue technological trends, eventually offering to the end-users devices and innovative technologies that could become a 'dead weight' during their cultural experiences. However, by means of innovative technological applications and location-based services, it is possible to shorten the distance between cultural spaces and their visitors (2015, p. 209).

Traditions and localization are, unarguably, the key authenticating definers of objective authenticity. Digital making of souvenirs reinforces a negotiated authentication process. For instance, Vettese, Anastasiadou, and Vones (2017) use the processes and experiences of making 3D printed souvenirs of place to investigate their potential applications with regard to modern 'overlooked' heritage. According to the authors: "the process of personalizing and printing out one's own souvenir on site adds potential for participation, interactivity and further engagement and added value to the visual and cultural heritage of the site. This process embeds the visitor's individuality and 'self' within the experience, object and place. It is anticipated that this memento will add to the highly personal nature of the bond between the individual, the object and the place by formulating an 'autobiographical' element" (Vettese et al. 2017, pp. S3683, S3684).

It is argued that digital creation through 3D printing holds the attention of the maker in a manner that varies from machine-made handicrafts (Rotman 2013). Two types of workmanship are described by Pye: the 'workmanship of risk' can be described as workmanship using any kind of technique or apparatus, in which the quality of the result is not determined. The 'workmanship of certainty' is that which can always be found in the quantity of the production. The quality of the result is always predetermined before a single saleable thing is made (1968, p. 4). Both risk and certainty are reflected in 3D printing. Gershenfeld explains that "digital fabrication desegregates 'hard and software and physical science from computer science'. The core theme of our research on digital fabrication is not about computers controlling tools, but about the computer itself as a tool" (2012, pp. 57–58).

Furthermore, single smart spaces can transform a cultural space into a smart cultural environment to enhance the enjoyment and satisfaction of the involved people. In support of 3D printing, Anastasiadou and Vettese opine that:

> mass-market production of souvenirs, their disposability and their mixed up, interpretive styling may detach the tourist from the actual experience. Conversely, it is the personal relationship that is formed between the tourist and the souvenir that makes the object authentic. The personalization of souvenirs, through 3D printing, offers opportunities for a different approach to manufacturing that influences notions of authenticity. In this way, it is possible to escape the serial reproduction of culture, engage tourists in the creation of meaning, and (re)frame the connections among them, their visited places, and their souvenirs (2018, p. 165).

With regard to the authentication process of ethnic handicrafts, Schilar and Keskitalo point out that, "when ethnicity is said to be manifest and practiced through handicrafts, these seemingly innocent objects become political. They raise questions concerning the authentication process such as: who can do what handicraft, who can use what symbols or what developments" are permitted (2018, p. 36). Ethnic norms and peripheries continue to be authenticated in a negotiated manner as tourism grows to be an important economic venue. Schilar and Keskitalo (2018) use a social constructivist approach to examine negotiated authentication of ethnic margins and boundary-making in handicrafts in northern Sweden, Norway, and Finland. The crafters either authenticate themselves as belonging to an ethnic community or through their artistic quality of work. Schilar and Keskitalo (2018) share that:

> Saminess is protected through their methods of protecting and sustaining Sami handicraft culture, the Dduodji organizations also use the blood boundary to keep Sami handicrafts exclusive – the authentic Duodji is made by Sami. Some Swedish and Sami artists speak about the exclusiveness of Sami handicraft as perhaps being 'unnecessary' or not particularly modern. They wonder whether a person born and raised in Stockholm should have a greater claim on Saminess than some other crafters. They understand that the norms of today have emerged

> from a specific history, but suggest that it might be ‘time to move on’, especially in the context of handicrafts. (2018, p. 42)

Tourists are on the other side of the mutual authentication equation. The manner in which they authenticate their trip phases and engage with their souvenirs upon return home also constitutes an interesting area of study. Some recent studies have examined authentication of souvenirs by consumers and offer insights on how they are existentially authenticated at home as “they represent distant places in people’s homely environment and encroach in their personal spaces. That is, how they mediate in tourist lives in their home. They perform a magic role in bringing distant, faraway places into the orbit of peoples’ everyday lives” (2017, p. 53).

Autoethnographic narratives can offer insights on how meanings are constructed and reconstructed when tourism traverses day-to-day life. Souvenirs make their entry and the selected ones “live a prolonged afterlife in the home and are not locked away or delegated to other uses” and the ones that can be domesticated ‘live on’ (Haldrup 2017, pp. 55, 57). Furthermore, Haldrup says that souvenirs influence the state of being and “an important part of their fascination lies in their ability to affect us, afford particular moods, sentiments and imaginations. They are living with us than for us. They offer promise, remembrance echoes, comfort, disturbance, hope and affection. In doing so they become co-constitutive of human life worlds and imaginations, actively interrupting and intervening it” (Haldrup 2017, pp. 57, 59).

Taking inspiration from the above, it is evident that existentialist authenticity shapes the manner souvenirs are experienced by tourists after they are brought home; but objectively authentic perceptions guide the selection and purchase ritual. Clearly, experiential consumption of souvenirs is enriched by self-authentication in the backdrop of home and nostalgic memories of the trip. Producers and retailers on the other hand, undergo their own complex authentication process to build, present, and sell the souvenirs. Digital innovations can enrich this process by continuous informing, updating, and authenticating the experiences with the ‘othered’ object. An immersive and interactive journey shaped by the heritage institutions, for instance at a historic site or a museum, reawakens in home settings upon return. A resilient mutual authentication process is merited that harmoniously optimizes the supplier resource/goals and emotionally authenticates consumer experience at different stages of travel and meaning making upon return.

Situating this discourse in the context of COVID-19, the souvenirs have taken a whole new meaning and as emblems of emotional expression and pleasant memories. Many cultural institutions are selling face coverings that feature artwork and cultural expressions to earn income. In reference to the new must-have museum souvenir, Marshall (2020) writes that masks have become a popular saleable commodity at museum gift shops. Cultural institutions are showcasing their artworks or logos on face coverings and selling them to earn revenue as they struggle to reopen. Masks with art designs and impressionist paintings have become a big draw across the world such as at Klimt Villa in Vienna, at the Metropolitan Museum of Art, and at the Rijksmuseum in Amsterdam (Marshall 2020). The pandemic era is a historic time and new history is being made. Unique utilitarian souvenirs can help commemorate the sociocultural essence of the COVID-19 environment. As Weisstouch (2020) aptly writes: These days, as a global pandemic has paused the world and rendered the simple act of going to buy a grocery item risky. Everyone is at an edge and cautious. During these times, the simple act of holding an object can reawaken sights and sounds of jubilant moments.

References

Altintzoglou, T., Heide, M., & Borch, T. (2016). Food souvenirs: Buying behaviour of tourists in Norway. *British Food Journal*, *118*(1), 119–131.

Anastasiadou, C., & Vettese, S. (2018). Digital revolution or plastic gimmick? Authenticity in 3D souvenirs. In *Authenticity & tourism: Materialities, perceptions, experiences* (pp. 165–179). Emerald Publishing Limited.

Asplet, M., & Cooper, M. (2000). Cultural designs in New Zealand souvenir clothing: The question of authenticity. *Tourism Management*, *21*, 307–312.

Birkeland, J. (2002). *Design for sustainability: A sourcebook for integrated, eco-logical solutions*. London: Earthscan Publications Ltd.

Cappucci, M. (2016). Indigenous tourism in the Amazon region of Suriname: Actions to preserve authenticity and natural resources. *GeoJournal of Tourism and Geosites*, *17*(1), 47–56.

Cave, J., & Buda, D. (2018). Souvenirs in dark tourism: Emotions and symbols. In *The Palgrave handbook of dark tourism studies* (pp. 707–726). London: Palgrave Macmillan.

Chhabra, D. (2005). Understanding authenticity and its determinants. *Journal of Travel Research*, *44*(1), 64–73.

Chhabra, D. (2010). *Sustainable marketing of cultural and heritage tourism*. London: Routledge.

Chhabra, D. (2015). *Strategic marketing in hospitality and tourism*. Nova Science Publishers, Inc.

Chhabra, D., Healy, R., & Sills, E. (2003). Staged authenticity and heritage tourism. *Annals of Tourism Research*, *30*, 702–719.

Chhabra, D., Sills, E., & Rea, P. (2002). Tourist expenditures at heritage festivals. *Event Management*, *7*(4), 221–230.

Chianese, A., Piccialli, F., & Valente, I. (2015). Smart environments and cultural heritage: A novel approach to create intelligent cultural spaces. *Journal of Location Based Services*, *9*(3), 209–234.

Ciftci, H., & Walker, S. (2017). Design for grassroots production in eastern Turkey through the revival of traditional handicrafts. *The Design Journal*, *20*(1), S2991–3004.

Cohen, E. (1993). Investigating tourist arts. *Annals of Tourism Research*, *20*(1), 1–8.

Cohen, E. (2000). *The commercialized crafts of Thailand*. Honolulu, HI: University of Hawaii Press.

Esperanza (2008). Outsourcing otherness: Crafting and marketing culture in the global handicrafts market. *Research in Economic Anthropology*, *28*, 71–95.

Feest, C. (1992). *Native arts of North America*. London: Thames and Hudson.

Flynn, A., & Flynn, V. (2012). *Custom nation: Why customization is the future of business and how to profit from it.* Dallas, TX: Benbella Book Inc.
Fu, Y., Liu, X., Wang, Y., & Chao, R. F. (2018). How experiential consumption moderates the effects of souvenir authenticity on behavioral intention through perceived value. *Tourism Management, 69*, 356–367.
Gershenfeld, N. (2005). *Fab: The coming revolution on your desktop – from personal computers to personal fabrication.* New York, NY: Basic Books.
Gordon, B. (1986a). Souvenirs of Niagara Falls: The significance of Indian whimsies. *New York History, 67*(4), 389.
Gordon, N. (1986b). The souvenir: Messenger of the extraordinary. *Journal of Popular Culture, 20*(3), 135–146.
Graburn, N. (1976). Introduction: Arts of the fourth world. In N. Graburn (Ed.), *Ethnic arts and tourist arts: Cultural expressions from the fourth world* (pp.1–32). Berkeley, CA: University of California Press.
Graburn, N. (1987). The evolution of tourists arts. *Annals of Tourism Research, 11*(3), 393–420.
Graburn, N. (1982). The dynamics of change in tourist arts. *Cultural Survival Quarterly, 6*(4), 7–11.
Haldrup, M. (2017). Souvenirs: Magical objects in everyday life. *Emotion, Space and Society, 22*, 52–60.
Halewood, C., & Hannam, K. (2001). Viking heritage tourism: Authenticity and commodification. *Annals of Tourism Research, 28*(3), 565–580.
Haro Zea, K. L., Haro-Zea, Y. R., Roblero-Mazariegos, G., & Sánchez Sánchez, S. (2018). Chiapaneca handicraft as a driver of sustainable local development. *Global Journal of Business Research, 12*(2), 73–81.
Hitchcock, M., & Teague, K. (Eds.). (2000). *Souvenirs: The material culture of tourism.* Sydney: Ashgate.
Hollinshead, K. (1996). Disney and commodity aesthetics: A critique of Fjellman's analysis of "Distory" and "Historicide" of the past. *Current Issues in Tourism, 1*(1), 58–97.
Kord, B., Heidari, Z., & Rigi, F. (2016). Identification, and ranking of marketing strategies in handicraft industries of Fars province Iran with AHP attitude. *IIOAB Journal, 7*, 397–404.
Littrell, M., Andersen, L., & Brown, P. (1993). What makes a craft souvenir authentic? *Annals of Tourism Research, 20*, 197–215.
Littrell, M. A. (1990). Symbolic significance of textile crafts for tourists. *Annals of Tourism Research, 17*(2), 228–245.
Luscombe, A., Walby, K., & Piché, J. (2018). Making punishment memorialization pay? Marketing, networks, and souvenirs at small penal history museums in Canada. *Journal of Hospitality & Tourism Research, 42*(3), 343–364.
Maghsoodi, S., & Nadalian, A. (2018). The interaction of "Globalization" and Persian "Handicrafts": An analytical investigation. *Journal of History Culture and Art Research, 7*(5), 123–132.
Marshall, A. (2020). The new Must-have Museum souvenir. Retrieved on June 15, 2020, from https://www.nytimes.com/2020/07/16/arts/design/museums-masks.html.
Morgan, N., & Pritchard, A. (2005). On souvenirs and metonymy: Narratives of memory, metaphor and materiality. *Tourist studies, 5*(1), 29–53.
Moscardo, G., & Pearce, P. (1999). Understanding ethnic tourists. *Annals of Tourism Research, 26*(2), 416–434.

Paraskevaidis, P., & Andriotis, K. (2015). Values of souvenirs as commodities. *Tourism Management*, *48*, 1–10.

Petrelli, D., Marshall, M. T., O'brien, S., McEntaggart, P., & Gwilt, I. (2017). Tangible data souvenirs as a bridge between a physical museum visit and online digital experience. *Personal and Ubiquitous Computing*, *21*(2), 281–295.

Popelka, C., & Littrell, M. (1991). Influence of tourism on handicraft evolution. *Annals of Tourism Research*, *18*(3), 392–413.

Pye, D. (1968). *The nature and art of workmanship*. Cambridge: Cambridge University Press.

Reville, G., & Dodd, T. (2003). Authenticity perceptions of Talavera pottery. *Journal of Travel Research*, *42*, 94–99.

Rotman, D. (2013). The difference between the makers and manufacturers. *Technology Review*, *116*(1), 76–79.

Schilar, H., & Keskitalo, E. C. H. (2018). Elephants in Norway: Meanings and authenticity of souvenirs from a seller/crafter perspective. *Tourism Culture & Communication*, *18*(2), 85–99.

Schilar, H., & Keskitalo, E. C. H. (2018). Ethnic boundaries and boundary-making in handicrafts: Examples from northern Norway, Sweden and Finland. *Acta Borealia*, *35*(1), 29–48.

Silberberg, T. (1995). Cultural tourism and business opportunities for museums and heritage sites. *Tourism Management*, *16*, 361–365.

Swanson, K. (2004). Tourists and Retailer's perceptions of souvenirs. *Journal of Vacation Marketing*, *10*(4), 363–377.

Thomsen, L. (2018). Retailing in places of world heritage, transition and 'planned authenticity'. *Geoforum*, *91*, 245–252.

Timothy, D. (2005). *Shopping tourism, retailing and leisure*. Clevedon: Channel View Publications.

Timothy, D. J. (2011). *Cultural heritage and tourism: An introduction*. Bristol: Channel View Publications.

Tsai, C., & Wang, Y. (2017). Experiential value in branding food tourism. *Journal of Destination Marketing & Management*, *6*(1), 56–65.

Vettese, S., Anastasiadou, C., & Vones, K. (2017). A study of the relationship between personalised 3D printed 'Souvenirs of Place' and public perception of modern architectural heritage. *The Design Journal*, *20*(sup1), S3683–S3695.

Weisstouch, L. (2020). In defense of souvenirs: For travel-lovers stuck at home, even the unlikeliest tchotchkes can spark joy. Retrieved June, 2020, from https://www.washingtonpost.com/lifestyle/travel/in-defense-of-souvenirs-to-travel-lovers-stuck-at-home-even-chintzy-objects-can-spark-joy/2020/04/16/a281783a-7da6-11ea-9040-68981f488eed_story.html.

Xie, P. F., & Wall, G. (2002). Visitors' perceptions of authenticity at cultural attractions in Hainan, China. *International Journal of Tourism Research*, *4*(5), 353–366.

12 Conceptualizing a smart resilient authentication system in transformative times

This chapter summarizes diverse authenticity positions in the context of sustainability, economic value, marketing, and the authentication process. Dynamics, behind the apparently rational appeal for negotiated authenticity, are discussed. This chapter also offers further deliberations on how the heritage tourism industry is reinventing itself and holds potential to purify/moralize authenticity and the authentication processes as it adapts/counters the impact of COVID-19.

Numerous scholars postulate that history floats, that is, it is not fixed. It is but one perspective of the truth. Past representations remain selective in nature because "history is distanced from personal or collective memory" (Chhabra 2008, p. 433). Reality cannot be reproduced completely; therefore, it is almost impossible to embrace it in an objective manner. Social and political changes shape human perceptions and observations which are limited by physical and cognitive restraints (Fu, Liu, Wang, & Chao 2018; García-Esparza 2016; Grau, Ginhoux, & Riera 2003; Meng, Cai, Day, Tang, Lu, et al. 2019). In other words, pure objectivism is impossible. For several decades, authenticity has been appropriated to concepts of reality, truth, identity, self enhancement, continuity, change, rationality, sincerity, and commodification that form the underpinnings of the leading authenticity theories today.

Undeniably, the objective form of authenticity remains the reference point in heritage tourism, but it continues to be negotiated to reverberate with the maker, the mediator, and/or the receiver.

All arguments lend credence to the evolving and relativized nature of authenticity as efforts are made to maintain continuity to some extent before the tradeoff point again shifts to embrace change and/or gratify the constantly reconstructed human self. In the Theseus's paradox: "Theseus's ship was kept by the Athenians as a memorial for a long time. Due to the gradual replacement of rotten planks, the ship retained its original form but its material was entirely renewed. The question was then raised: was it still Theseus's ship, thus retaining a certain identity. On the other hand, one could suggest that the materials that were removed could have been reassembled" (Markovic 2018, p. 12).

Looking at the diverse authenticity positions in the context of sustainability, economic value, marketing, and the role of different authentication agents and the authentication process, this book presents critical insights on the dynamics behind the apparently rational appeal for negotiated authenticity. In Chapter 2, I pinpoint the need for ongoing resilience strategies to protect the vulnerable nature of negotiated authenticity so that the core or original essence is retained in the face of inevitable change. Undeniably, resilient communities rely on digital interventions, more acutely so, in the pandemic times. But equally important is the strategic and controlled use of technology to promote direct physical connection and experiences in time and place (Syvertsen & Enli 2019, p. 2). Noticeably in an effort to remain competitive, many more heritage settings, across the globe, are employing digital media of some sort (Hand 2016). Numerous studies from the pre-pandemic era also lend support to the digital detox standpoint. But, during intra-pandemic times, technology has played a crucial role in connecting people with each other and to the outside world and places of interest. The global pandemic has brought a sudden pause to various aspects of life and closed the doors of heritage attractions and other related leisure/recreation spaces due to mandatory restrictions and social distancing regulations imposed by governments across the globe (Anderson & Knee 2020). During these unusual turning points in humanity, the digital medium has informed/entertained/charmed people and performed a powerful role in transmitting messages/heritage stories/programs from heritage institutions such as museums, art centers, and historic sites. Therefore, the pause in heritage tourism and the socioeconomic destruction caused by the pandemic compels for novel insights on resilient techniques to revive and/or reform heritage experiences, keeping the new normal in mind. Toward the end of this chapter, I will offer further deliberations on how the heritage tourism industry is reinventing itself and how it can purify/moralize authenticity and the authentication processes as it adapts or counteracts the impact of COVID-19 on heritage tourism.

In this book, I have added weight to the notion of shared or collected authentication by lifestyle entrepreneurs (as in the case of homestay tourism) and entrepreneurial heritage institutions rooted in localized agenda. The present use of past cannot be neglected because the audience is living in the present and interpreting the past in contemporary times and way of life. Therefore, authentication of negotiated authenticity can delicately endeavor to make the past meaningful to the contemporary audience, while sustaining its core essence for future generations. In Chapter 3, stress is on a cohered approach or genuinely negotiated standpoint that deeply scrutinizes the authentication process, using political and socio-constructivist benchmarks. Authentication strategies should move beyond the political systems and promote localized voices in traditional objects and experiences; this calls for the embrace of impartial procedures to nurture folklore and local histories.

In the context of homestay tourism (Chapter 7), I discuss the role of lifestyle entrepreneurs as they negotiate objective authenticity, in selective showcasing and sharing of their homes and traditions. I point out that a lifestyle entrepreneur's vulnerability is his or her ability to anticipate, compromise, cope, and recover from shocks over an extended period of time and is dependent on access and claim to related sociopolitical, economic, and environmental resources and power systems. In other words, a government-induced support system needs to be in place to promote these entrepreneurs. A sustainable livelihood (SL) approach is suggested that is forward centric, collaborative, and integrates voices of indigenous, especially remote, populations to authenticate and reform rural tourism. In the context of nation branding (Chapter 8), I reiterate the need to authenticate pluralism in a harmonious manner to melt dissonance and conflicting representations. The aim is to make heritage resilient by affording voice to both silent (by appropriating the presence of the noticeably or quietly 'absent') and vocal communities so that they can enrich an exotic 'othered' image of a nation. Alternative routes are required that can root the objectively authentic essence in the marketing messages. In addition to this, the nations have to prove their resilience through their ability and capacity to strategically counter crisis situations. The COVID-19 has laid bare the economically and politically vulnerable and weak aspects of nations across the globe in addition to exposing the fallacies of existing health and well-being systems. In Chapter 10, I open the debate on the role of museums as authenticating agents of culture and heritage (they host, showcase, and make authenticity) and discuss the manner in which this role shapes and confirms authority. Furthermore, these institutions are striving to contextualize the history and pastness of heritage to recontextualize them and make meaning in troubled times so that they can offer resonating deep experiences on digital platforms. In Chapter 11, I turn attention to ethnic restaurants and offer insights on surging (but currently on hold due to social distance measures) demand for entertainment experiences in 'othered' settings rendered objectively authentic with strategic negotiation and meticulous selection of authenticating markers.

By offering a variety of heritage settings and manifested scenarios, this book offers dialogical insights on how authenticity is pursued and authenticated. Numerous heritage agencies and their expressions are examined. Although each setting is unique with its own particularities, all homogenously attest to the dynamic nature of authenticity and its routing through complex negotiated authentication processes. Some heritage agencies (such as museums, ethnic restaurants, and lifestyle entrepreneurs of homestays in remote settings) are more advanced in setting up clear objective benchmarks, whereas others such as artisans and artists from ethnic communities shift their traditional infusions into heritage tourism products and experiences like a pendulum, negotiating between change and continuity. For instance, much artwork lends itself to meta-credence, implying it is subject to a negotiated authentication process because its complete originality can never be traced and/or verified (Ekelund, Higgins, & Jackson): "the key to authentication in the art market is the establishment of credence in the genuine character of the piece of art traded. Naturally, living artists can attest to work done by their hand. Most goods carry a probability of verification but some of these goods entail an extreme or zero probability of ever being verified. The agreement of experts determines credence status where expert consensus is substituted for falsification. The quality or authenticity of credence goods can be known but a meta-credence good's authenticity may never be known or, if discoverable, only at an extreme cost" (2019, pp. 1–2).

In other words, most heritage and works of art are meta-credible by nature as it is not possible to totally confirm their authentic quality; a rational appraisal of probabilities is required and the expert might deliberately label a picture piece of art as fake to avoid financial liability. On the contrary, even when authenticity cannot be verified, an expert might sway in favor of clients and market demand and deem a fake of work of art as authentic. In both cases, a strategic negotiation strategy is pursued to resonate with the personal agenda of the authoritarian that can be shaped by macroenvironment forces such as economic, social, political, and natural (Ekelund et al. 2019).

By the same token, traditional craftspeople also practice a domestication strategy to navigate and decommercialize traditional crafts (Holroyd, Cassidy, Evens, Gifford, & Walker 2015). Holyroyd et al. note that: "the most frequently employed design strategy is the redesign of craft products to suit the needs of contemporary customers. Meanwhile, the economics of making items for sale places pressure on the time that can be spent on each item leading to the 'trinketisation' of crafts" (2013, p. 3). In this context, Cohen (1989) argues that negotiation is unavoidable as all traditional crafts evolve with time and are subject to fluctuating social, economic, and political arrangements. Stankard (2010) posits that such changes are an intrinsic and unavoidable characteristic of traditional cultures. To maintain some sense of continuity, Holyroyd et al. share a locally cultivated scheme focused on involving the amateurs who can give time and care and use skill and creativity, while retaining the intrinsic qualities of the crafts. The authors write:

> the traditions must evolve in order to remain relevant and viable in contemporary life. It is important for practitioners of traditional crafts to have space to experiment and innovate. However, it is important to note that fixed strategies do not necessarily 'lock in' makers to prescribed patterns. As makers gain confidence and develop their tacit knowledge, they often begin to adapt and deviate from patterns, and to seek out resources that will support them in doing so. Therefore, as with skills, we can see that the domestication strategies discussed here offer a means of progression to amateur practitioners. (Holroyd et al. 2015, p. 9)

In Chapter 11, I pinpoint popular demand for authentic experiential consumption of souvenirs and also offer insights on how suppliers use numerous innovative authentication techniques to lend credence to heritage merchandize and optimize deep connections. Also, I discuss the mutual authentication process through which souvenirs tangibilize visitor experience both, at the site and upon return home, by reviving nostalgic memories. I argue that mutual authentication helps to co-create an experience by optimizing the self in the presence of an othered setting or object. Inferences drawn here support a harmonious synergy between the subjective and the objective nature of authenticity. Aligned with the pandemic times, evidence can be noted of the manner masks are being designed with artistic or heritage imprints to prompt consumer attention and connection with the actual piece of heritage or art (Weisstouch 2020).

Furthermore, it is important to note that the process of authenticating and selling handicrafts initiates from a critical juncture where cultural pride harmoniously carves the route for economic returns (Heller, Pietikäinen, & da Silva 2017). Capitalism and global political-economic environments have generated conditions that "(re)structure traditionalist and modernist discourses about artisans and their historical bodies and their connections to the local land (nature) and how they interactionally authenticate and sell their products" (Heller et al. 2017, p. 2). Heller et al. further observe that:

> While things play out slightly differently for each artisan, nonetheless, together they point to some important shifts in the idea of what handicrafts may be, in particular, as emblems of an organic nation. The value of the product is understood not as use-value but as a symbolic added value in a competitive market and as a monetized value in economic development. The question remains how they will be transformed, or whether the strategies adopted by people like the artisans examined here sufficiently neutralize tensions to allow the contradiction to be socially, politically and economically productive. (2017, p. 34)

Subjectivities in the name of art, entrepreneur, economic developer, citizen of the world, and collective politicism have gradually permeated the advocacy spheres of marginalized cultures. Coupled with this, has been the emerging demand for immersion into the traditions of minority cultures. Immersive stance echoes the core essence of the existentialist authenticity theory that advocates self-making by offering a window to connect with one's inner self (self-making) and with others (Brown 2013). While elaborating Sartre's view of self-authenticity, Daigle explains:

- Self-authenticity must constantly be reclaimed as one continues in this life.

To be authentic, one must hold oneself thoughtfully to embrace one's world and one's relationships with the others. In other words, "I could lose myself in order to recuperate myself as an enriched being made through experiencing myself as being with-others. In fact, engaging in such relations becomes essential to the process of 'distantiation' – the key of authentic becoming;"

- The self makes itself and is never static. One must actively engage in the formation of the self. Human, being wordly beings, are subject to power relations and the subject of their own relations to themselves; and
- To become active subjects of our own lives calls for embracing our fluid and evolving state and recognizing the 'distantiation' process that forms the core of our being. It also requires us to acknowledge that we will always unwrap historically and through our relationships and experiences. In other words, "our existence will always precede and supersede essence" (Daigle 2017, pp. 57–65).

Experience optimization, to the point of immersion, can be molded with the assistance of information communication technology (ICT) to develop hybrid innovative mechanisms. Authenticity is a culturally constructed quality and "is continually subjected to tremors and revisions because fluidity is inherent in the phenomenon as culture continues to evolve in the face of triggers such as changing micro and macro environmental factors" (Chhabra 2019, p. 6). Undoubtedly, its position will be impacted as it sorts itself out through the socioeconomic and wellness/health disruptions. Tourism is currently in a disequilibrium state and has halted. However, its resilience has been proved in previous disaster times and this time need not be any different. The tourism industry continues to equip and innovate itself with digital tools and in the absence of real audience, the heritage institutions have shifted their attention toward digital showcasing and immersion to connect with a broad spectrum of people looking for diversion and relaxation. Earlier, most agencies embraced digitalization to be competitive. However, in the intra-pandemic times, technology has become a key survival tool.

Today, creative and intelligent/sophisticated use of technology can help optimize visibility/remembrance in the minds of targeted heritage tourism markets. Deep probing of layers of authenticity of the individual self and that of both tangible and intangible heritage expressions can be unfolded and shared in a meaningful manner. Multiple authenticities can also be presented and sought simultaneously, for instance, for tourists and pilgrims interested in visiting the Old City of Jerusalem the most popular monument, the Temple, is nonexistent. There are no remains and its original location cannot be accessed. In such cases, how can the site while celebrating the memory of the Temple showcase itself as authentic and actual? How can authenticity be constructed and what sources/authorities will legitimize it? Cohen-Aharoni discusses ways in which heritage sites adjacent to the Temple Mount construct claims of authenticity through adjusted guided tour performance: "by comparing three heritage sites that deal with the memory of the Temple: the Western Wall Heritage Tunnels (hereafter: the Tunnels), the Temple Institute, and the Archaeological Park of Jerusalem – Davidson Center, where claims to authenticity are advanced in spite of two challenges: the clear absence of the Temple itself, and the slight, yet important distance between the original location of the Temple and the locations of the heritage sites" (Cohen-Aharoni 2017, pp. 75–80).

Numerous heritage tourism settings have accentuated the promotion of mutual authentication between guests and hosts and/or heritage agencies to optimize experiential consumption. As pointed out earlier, heritage tourists and suppliers of heritage offerings mutually authenticate experiences. Authenticity has, undoubtedly, become an important experiential consumption tool. It is closely connected with 'a sense of self and identity' and is a leading motivation factor and generator of satisfactory consumer experiences (Fu et al. 2018). In the contemporary era, it is mostly about establishing emotional bonds with the physical settings and objects and seeking optimized experience; heritage tourists especially appear to seek experiences that help them to get in touch with their true selves (Breathnach, 2006; Poria, Biran, & Reichel 2006).

Domínguez-Quintero, González-Rodríguez, and Roldán (2018) look at the usefulness of different authenticity domains in the context of experience quality and emotional connections. The authors find that objective and existentialist authenticities have a positive impact on experience, stimulate positive emotions, and elevate satisfaction levels. Meng et al. (2019) focus on the motivational aspects of authenticity when they examine the individual well-being urban migrants from rural areas in China to gather insights on the influence of nostalgia and perceived authenticity. The authors find that 'old home' memories generate nostalgic motivations that inspire rural area visits by urban migrants. Consumer in the contemporary era desires emotionally charged negotiated authentic experiences (Pine & Gilmore 1999). An authentic experience produces stories waiting to be told (in the form of WOM) so that it continues to breathe in memories. It is of no surprise then that the consumer yearning for optimized immersive experiences has both fueled and shaped the demand for innovative (especially digital) interventions.

Demand for experiential consumption calls for object authenticity to be expressed, showcased/experienced in a manner that prompts emotional connections. Therefore, "with the increase in consumerism and the edutainment culture required for the heritage field to flourish, the main question for the survival of heritage museums in the 21st century is whether curators can recognize and integrate other forms of representation to fit the object into the minds of present-day multi-perspective audiences. As museums are not isolated from the rest of the world, they have to reflect the current culture of the digital age in order not only to preserve, but also to promote artefacts" (Markovic 2018, p. 8). These innovative digital tools are heightening the resilience of heritage tourism during the intra-pandemic and should be useful in the post-pandemic times. Notably, more so now, heritage agencies are adapting to optimize opportunities offered by the ICT and making their repositories meaningful in the pandemic times. For instance, the Phoenix Art Museum has taken a virtual initiative to connect their collections with different faith groups:

> In a world of more than 7 billion people, innumerable beliefs define our creeds, our identities, and our cultures. For the billions of us who have lived throughout history, we've learned that religious belief has the power to transform entire nations, to unite and divide, to conquer and to heal. While our differences in beliefs can at times polarize us, faith traditions have also influenced and inspired the creation of some of the most compelling works of art in our collective history.
>
> For this week's virtual visit, we explore works in the PhxArt collection inspired by some of the major faith traditions in the world, from Christianity to Islam, Hinduism to Judaism, and so much more. Join us to discover the ways belief can come to life in a range of media, take a deep look at the Museum's collection and past exhibitions, and enjoy belief-inspired playlists, reading lists, and more (Phoenix Art Museum 2020).

Almost a decade ago, Jantz pointed out: "many institutions are hesitant to incur the overhead implied by the processes. However, undergoing the certification process and providing a new service – creation of authentic digital objects – is not only necessary to support the integrity of scholarship, it also offers the opportunity to extend traditional roles into the digital environment" (2009, p. 80). In Chapter 4, I have offered insightful views on how heritage agencies are embracing ICT. Moving forward, I would like to elevate the role of digital surrogates in fortifying resilient authentication systems, especially in times of disequilibrium. Recent literature posits that the digital future of heritage is leaning toward digital surrogates, digital way of conserving in an enduring manner, and the democratization of technology to enable easy access to digital resources and plan out digital workflows. Democratization of technology implies simplifying, reducing costs, easy to use, and harmonious with prevailing cultural environments.

In the context of digital surrogates, Mudge, Ashley, and Schroer (2007) discuss how democratization of technology can support their successful development and presentation (see Table 12.1). The authors describe the manner in which scientifically tested attribution can confirm authenticity and reliability of digital surrogates, while easy access can ensure availability for present and future generations. In support of digital surrogates, Mudge, Ashley and Schroer write:

> Digital surrogates of our 'real world' cultural heritage can robustly communicate the empirical features of CH (cultural/heritage) materials.

Table 12.1 Digital surrogates

Digital surrogates are the undistorted and only existing versions of original heritage that might have been lost or wiped out (ruined). They make possible scientific scrutiny and personal delight without the presence of the actual object or place. To be true to their function, they have to be replica of the original by apprehending the minutest details.
Digital surrogate is a kind of commodified presentation of the actual artefact, serves as an educational and promotional tool due to its digital presence (Stanco, Battiato, & Gallo 2011) and can be disseminated online and made accessible to the scholarly community. This enables sharing knowledge about the artefact and developing pooled interpretations and popular reenactments of the past.
Additionally, a digital copy can forge a close connection with the artefact and offer a complete view from all angles/positions as it is transformed into a digital 3D platform. An important vantage point is that it offers interactivity in real-time and therefore can hold audience attention for an extended period of time.
Another key benefit is that a digital surrogate is the insurance the digitalization provides. In the event of a natural, accidental, or manmade crisis, if the actual artefact is destroyed, it remains available in a digital form for the purpose of research as well as entertainment.

Source: Markovic (2018)

> When digital surrogates are built transparently (that is, according to established scientific principles), authentic and reliable scientific representations can be developed. These representations allow repurposing of previously collected information and enable collaborative distributed scholarship. Digital surrogate archives remove physical barriers to scholarly and public access and foster widespread knowledge and enjoyment of nature and our ancestors' achievements. (2007, p. 1)

The systematic extraction of information from digital pictures can help construct digital surrogates describing the 2D and 3D positions, their setting, content, and substance. New technologies include "single and multi-view reflection transformation imaging, the algorithmic extraction of surface feature drawings from reflection information, as well as photogrammetric breakthroughs that permit automatically calibrated and post-processed textured 3D geometric digital surrogates of objects and sites" (Mudge et al. 2007, p. 5).

The emergence of vigorous digital biopic instruments offers mechanical (robot-like) management of the post-procurement process, and the automation requires minor configuration with potential to enhance reliability and enormously reduce the computer technology expertise necessary to manage a digital workflow (Mudge et al. 2007). These methods optimize ease of use of new knowledge to assist CH professionals in building digital surrogates and cut short their time investment so that they are able to learn other skills such as digital photography that can be prove to be cost effective in embracing digitalization. Mudge et al. further write that:

> Rich 2D and 3D information can be captured with the equipment found in a modern wedding photographer's kit Reflection Transformation Imaging (RTI); invented by Tom Malzbender of Hewlett-Packard Laboratories (HP Labs), it is an example of computational extraction of 3D information from a sequence of digital photographs. RTI data acquisition analyzes reflections from a subject's surface. When a surface is photographed from a fixed position and illuminated from different known locations, its properties of shape and many material attributes (including color) can be computationally revealed. Reflections disclose shape by capturing the directional vector, mathematically named a 'normal' that is perpendicular to the surface at the photographically sampled location. Knowledge of surface normals permits construction of the surface's 3D geometry as in the process of photometric stereo or codification of the normal information on a per-pixel-basis in a 2D image as in polynomial texture mapping (PTM) (Malzbender, Gelb, & Wolters 2001). 3D lighting models use this normal information to permit relighting of the subject from any direction and with any illumination source in interactive viewing software. RTI has been widely used in law enforcement, natural science, and cultural heritage. (Mudge et al. 2007, p. 5)

Markovic (2018) offers several examples of how heritage sites have become vulnerable during times of natural calamities, wars, vandalism, or normal wear and tear. Numerous digital initiatives are now focusing on reviving these damaged sites or developing surrogates to minimize their use. The author posits that:

> the focus of heritage preservation today should be not only on the maintenance of the materiality of an artefact as a physical substance present in the real world, but also on developing workflows for providing 'digital materiality' for that artefact; in other words, creating an authentic digital copy before the original is lost forever. The continuous development of new sensors, data capture methodologies and the improvement of existing ones contribute significantly to the 3D documentation, conservation, and digital presentation of heritage and to ... scholarship. The innovations and improvements in digital 3D scanning devices as well as increasing computational powers have provided new means for generating a high-quality digital copy of a heritage site or an artefact, which can be further used in photorealistic reconstructions, simulations or immersive VR experiences for educational and scholarly purposes. (Markovic 2018, p. 2)

Table 12.1 presents the advantages of digital surrogates in their ability to epitomize the 'original' content digitally in a trustworthy manner.

Clearly, digital surrogates represent a negotiated version of authenticity and are authenticated with a variety of tools. These digital wonders are proving to be useful as the heritage industry struggles to attract audience attention, especially during the intra-COVID-19 times, as they explore the prospects of stimulating virtual experiences and artificial intelligence, especially in the context of robotics and humanoids. Authentication test is the reliability exam and "while an exact copy of a 'real-world' artefact is a requirement for a digital surrogate, in Guidelines for the Preservation of Digital Heritage, it is argued that 'it may not be practical to expect an entirely objective guarantee of authenticity – there may always be an element of trust or subjective judgment in deciding that authenticity has been sufficiently proven" (Guidelines for the Preservation of Digital Heritage, UNESCO 2003). It can also be emphasized that the relationships, on which the required level of authenticity rests, must be thoroughly documented. Authenticity of a digital copy relies on being able to trust both the identity of an object – that it is, what it says it is, and the integrity of the object – that it has not been changed in ways that can alter its meaning. Maintenance of both identity and integrity also implies thorough documentation of the development process (UNESCO's Guidelines for the Preservation of Digital Heritage) (Markovic 2018, p. 11).

It is important to point out that although ICT offers innovative opportunities, it is not without fallacies. For instance, information crowding has become a constraint. Directional marketing of enterprises with clout, by dominant search engines and travel intermediaries, overshadows the existence of micro enterprises and handicrafts produced by marginalized/indigenous populations. Flaws, associated with robotic services at a heritage hotel, are shared in Chapter 6. From a smart heritage tourism marketing standpoint, existing resources need to be showcased in an equitable manner, optimizing economic benefits for marginalized and host communities. Smart search platforms are required to shadow the monopolistic presence of 'Giant' search engines such as Google, Yahoo, and MSN. Therefore, despite plethora of opportunities for cultural and heritage institutions, a large question mark continues to loom over the objective authenticity of digital heritage. From a consumer perspective, a balance between virtual and physical environments is vital. That said, it is important to remind that physical travel and onsite experiences are restricted, in the ongoing pandemic times, especially to enclosed heritage spaces such as museums and heritage sites. Some disconnect from online media and ICT can still be achieved in outdoor settings; that is, digital detox of heritage experience can happen if self-authentication is pursued, for instance, by holding a souvenir or a piece of art or some physical expressions of authenticity such as a hand-embroidered scarf (Weisstouch 2020). According to Syvertsen and Enli, "arguments for digital detox rest more on a presumption of balance that is akin to mindfulness; temporary breaks are seen as a vehicle to heighten consciousness and learn of self-regulation to reduce stress and increase the presence of here and now. A more authentic life is attainable through taking a break from digital technology, and that life improves with a more digital diet" (2019, pp. 2, 6). The authors build a case for the enhancement of authentic selves by periodically disconnecting from online media:

- First, there is the notion of authenticity as being real, genuine, and true, as opposed to being fake, unreal, and untrustworthy. In the digital detox context, there is an underlying argument that online communication is less true than offline communication, and that copresence in physical space is necessary for genuine and trustworthy communication. There is also the notion that online communication pollutes your mind and gives you a less than healthy body image. Hence, recommendations flourish to reduce the amount of online and social media communication and instead build trust and engage in truthful and trustworthy relations with people around you;
- The second notion of authenticity, as being true to one's inner values, comes across strongly in that one can regain back sense of time and space where you are grounded in a sense of 'hereness' and belonging. The aim is to reconnect with the sense of receiving more internal guiding as to how to live your life; and
- The third notion of authenticity is about a longing for a better yesterday, a more authentic era. In other words, a previous era is described as a less complicated time where life was slower and more down-to-earth, with fewer media temptations, and more direct contact with physical (rural) spaces. This life is described as better for body and soul (Syvertsen & Enli 2019, p. 12).

Therefore, digital detox heritage moments or programs can be facilitated, even during the crisis times. Some tourism businesses and organizations are using themes such as 'freedom from the mundane and home' to entice people to patronize their services. However, they have to follow, for instance the guidelines of the Center of Disease Control (CDC) in the United States (such as – run at 50% capacity) or face the risk of closure. Digital detox moments are healthy time off, to connect with one's self and become mindful of the physical surroundings. These meaningful connections have become crucial for the physiological and psychological well-being of people, especially in the pandemic times. They facilitate retrospection and authentication of the self, thereby opening inner consciousness toward moral behavior. In fact, hot authentication (mentioned in Chapter 3) can be referred as a self-authentication process that can strengthen self-identity; it holds potential to revitalize a sense of liveliness by inculcating a sense of 'hereness' and by targeting existentialist authenticity, both physiologically and in an effective manner (Hopper, Costley, & Friend 2015). Unarguably, existentialist authenticity is intrinsically tied with self-identity; and its negotiated dimension (theoplacity) is mostly about establishing emotional bonds in and with physical settings and objects and seeking optimized transformative experiences. For instance, holding a souvenir can evoke relaxation and a feeling of emotional connection with the object and the self through revival of memories.

At this juncture, I would like to draw attention to the notion of inauthenticity. Mkono (2020) refers to 'inauthenticity' as hypocrisy; that is, being inauthentic to oneself is akin to falling below the standards one sets for oneself. Moral capital is, many times, fakely claimed by using labels such as 'ethical tourists' because these people usually fail to live up to the moral standards they advocate. Linking hypocrisy with inauthenticity, Boaks and Levine posit that "whether by deception or hypocrisy, inauthentic people apply different moral standards to themselves that they do not extend to others" (2015, p. 164). How to be 'yourself?' has been examined extensively but inauthenticity is an emerging area of research (Mkono 2020) thereby prompting self-inquiry of one's proclamations and actual actions. Moral hypocrisy is deemed inauthentic because of dissonance between a publicly expressed belief and actual personal behavior (McDermott, Schwartz, & Vallejo 2015). Moral virtue is about putting others on a pedestal and situating one's own self on a pedestal. Virtue signaling is prevalent and is easy to detect, especially in the case of celebrities who often publicly make empty gestures, because of social media (Bartholomew 2015). A person who engages in virtue signaling can be deemed inauthentic in an existentialist sense (Wang 1999). Inauthenticity can take a different meaning for different dimensions of authenticity. For instance, in the context of objective authenticity, it refers to extreme commodification or commercialization of heritage. It becomes more complicated in the context of existentialist authenticity. What does it mean to be inauthentic in an existentialist sense? (Reiner & Steiner 2006). A person becomes unhealthy or unstable when he or she distances from his or her true nature and violates alignment with one's values, internal state, and identity (Gino, Kouchaki, & Galinsky 2015). Gino et al. (2015) suggest that feelings of inauthenticity leads to a lower sense of moral self-regard and impurity.

Existential authenticity relates to being true and authentic to oneself (Steiner & Reisinger 2006; Wang 1999). In fact, a person's mental well-being and self-actualization state is reliant on achieving existentialist authenticity – an optimal communion with the inner self. It is the alignment of daily behavior and the real self. And this state is not fixed but fluid and in a state of ongoing self-negotiation that scrutinizes accountability of personal thoughts, emotions, needs, and desires, and aligning one's actions with those experiences (Gino et al. 2015). The congruency between one's everyday behavior and one's 'real' self forms the basis for evaluating the existential authenticity we seek in our lives (Rinder & Campbell 1952). The sustainable outcome of pursuing existentialist authenticity at heritage sites has received scant academic attention. Existentialist authenticity can become inauthentic if 'truth of the moment' experiences are not lined up with the moral image one showcases outwardly. The consequences of inauthenticity can include a sense of alienation (rift from one's inner self), inducing in turn, a sense of purposelessness. According to Mkono, the pre-pandemic and the intra-pandemic times are "particularly antithetical to the retention of authenticity—with a large proportion of our existence ... spent and shared in the digital world, where the boundaries between real and fake, personal and social, private and public, have become more blurry than ever. An eco-conscious tourist must deal with all of these pressures and realities" (2020, p. 3). However, authenticness of digital heritage has been supported by several studies. It is also argued that truth of the moment experiences and existential immersions can take place via digital dais such as virtual reality. To what extent these immersions promote sustainable use of heritage resources is a question that requires probing. Moral selving can tone down quest for self-centered hedonic experiences. Rising global consciousness for a sustainable planet can strengthen a sense of morality which, in turn, can then harmonize hedonistic and spiritual quests.

Mkono (2020) discusses the inauthenticity dilemma in the context of ecotourism scenarios where tourists advocate themselves as environmentally accountable but fly to a pristine fragile area, intrude, and return. However, it is important to remember that ecotourism experiences have a strong heritage element in them. For instance, backpackers use homestays and immerse themselves in the culture of host communities. Aligned with an environmentally conscious tourist, a self-authentic or existentially authentic tourist claims to be genuine and seeks optimal state of exhilaration and oneness within a heritage setting or with a heritage object. Existentialist authenticity, in a way promotes, self-gratification at the cost of objective authenticity although one might argue that a self-authentic tourist in a heritage setting might generate lower negative impacts than a tourist who is out and about soaking the heritage ambience of the setting, instead of her or her own self. Hence, moral existentialism or morally driven theoplacity can alter self-gratification behavior and generate authentically conscious tourists.

Undeniably, authenticity has been predominantly used to describe the tourist quest for untouched and exotic cultures and the desire for immersion to escape from the mundane life. Existentialist authenticity has shifted focus toward the inner self in harmony with the external environment, to enjoy 'truth of the moment' experiences. Ethical (moral) consumption processes should examine the potential of optimal negotiation of existentialist authenticity to align with the sustainable transformation of one's optimal state of relaxation, one which is guided by responsible behavior. In other words, ethical consumption can both guide and support the moral selving process and diminish the sense of inauthenticity.

Hybrid use of digital and non-digital resources are needed to promote ethical production and ethical consumption practices. The pandemic has collapsed the global supply chains (Hall, Scott, & Gössling 2020). Heritage institutions pursuing ethical production strategies are investing in products and experiences that can build strong localized value chains. Conflicts between different role players in the heritage tourism value chain have intensified, thereby fostering the need to strategize inter-organizational governance (Chhabra 2015; Song, Liu, & Chen 2012). Song et al. (2012) developed a blueprint to facilitate efficient value chain governance in the heritage tourism system:

1. Examination of the effects of legal, economic, cultural, and other contextual factors on governance issues is important. Attention needs to be centered on integrating the efforts of tourism distribution channels and sustainable development and management of the value chain as a whole. Issues associated with power symmetry need to be addressed;
2. Comparison is needed between different tourism distribution channels, particularly suppliers and intermediaries, tour operators and travel agencies, and identification of issues associated with integration initiatives. The roles of tourists, in co-creating value, need to be explored;
3. Examination of the preconditions for, and effects of, different governance mechanisms, as well as the evaluation of outcomes, is still limited; and
4. To date, isolated research exists on the governance environment, structure, mechanisms, and outcomes in the tourism value chain. Their interrelationships, from an integrated standpoint, need to be investigated (Chhabra 2015, p. 244).

In this regard, a co-management approach (such as the one premised on reciprocal altruism) can guide relationships and collaborations between various partners/stakeholders in the heritage tourism supply chain for an all-inclusive negotiated/mutual authentication agenda. Reciprocal altruism resonates with effective co-management within and between various stakeholders whose support/cooperation is crucial to the heritage tourism organizations/enterprises (Chhabra 2015).

Trivers's Theory of Reciprocal Altruism offers a biological justification and postulates that altruistic acts are traded where anticipated benefits exceed costs. Synergies, guided by ethical principles, are required that use technology to develop a smart marketing agenda. The aim is to stay resilient in times of disequilibrium. The fundamental tenets of Trivers theory (1985, 2006) are:

a. there is a mutual relationship between helping others and helping oneself;
b. this cooperative relationship has a lag time;
c. benefits need to be larger than costs… net gain is perceived/needed; and
d. infrequent interaction can lead to cheating.

Furthermore, as pointed out in Chapter 5, smart marketing strategies hold potential to merge intelligent online and traditional communication mix efforts by innovatively and harmoniously optimizing both online and off-line presence. The fundamental aim is to negotiate the authentication process in a manner that maximizes benefits for different heritage agencies, consumers, and host communities while promoting moral selving, objective authenticity and sustainable use of heritage resources. Clearly, a smart resilient negotiated authentication system has to be multifaceted and needs numerous role players and harmonious integration of factors such as technology, digital detox, leadership/governance, community partnerships, stakeholder co-creations, negotiated authenticity, ethical consumption and ethical production, economic viability, and externality checks. In the forthcoming section, I elaborate on the fundamental tenets of resilience and propose a resilient negotiated authentication system for heritage tourism.

Resilience and negotiated authentication

The resilience notion is predominantly premised on the following fundamental principles:

1 the systems do not evolve in a linear fashion but according to a cycle, or loop;
2 the phases of the cycle roughly repeat but the characteristics of each stage, at different iterations, are not necessarily identical;
3 the characteristics and the speed of recovery from a destabilizing event depend on the system's adaptive capacity; and
4 the adaptive capacity depends on various forms of capital accumulated during previous phases and iterations (Cochrane 2010, p. 1).

Vulnerability refers to the ability to adapt. Resilience is a multifaceted phenomenon because it "takes account of turbulence and disequilibrium, self organisation and co-evolution" (Stevenson, Airey, & Miller 2009, p. 23). Chaos theory is also of relevance here as it helps to recognize that "large systems are inherently unstable and may be altered by a sudden trigger or stress event" (Cochrane 2010, p. 1). Fluctuations happen inside valleys or spheres of attraction and the system can self-organize on its own if disturbance happens at a point where the systems are strongest at the bottom or at the margins (Gunderson 2000). Therefore, paying attention to different steps of the resilient cycle (Table 12.2) can guide formulation of proactive strategies at each stage. For instance, strategic intervention strategies can be devised by drawing on different resources, depending on the vulnerability and brevity of each stage.

A resilient tourism system should include:

- Awareness of market factors and capacity to engage with them in an effective manner to support the 'triple bottom line' of sustainability. The 'triple bottom line' represents the environmental, social, and economic sustainability.

Table 12.2 Resilience cycle or the Holling loop

- *Reorganization* – rapid change after a destabilizing event, with regeneration and renewal of societal structures.
- *Exploitation* – new systems are created through exploitation of social and other forms of potential produced in previous phases. New institutions may emerge and new political, cultural, and social relationships form more easily.
- *Conservation* – the gradual construction of a new stable state, when structures are institutionalized and new capital is formed. Structures can become increasingly interconnected, leading to rigidity and inflexibility.
- *Release* – a disturbance event (or series of events) that destabilizes existing systems, releasing the rigidity of structures, and leading to the rapid changes.

Source: Holling (2001, p. 394)

- Collaboration and effective partnerships between stakeholders to enable equitable use and distribution of resources.
- Strong and consistent leadership guided by strategic and clear vision and effective management skills to address resource conflict and ability to effectively engage and unite the stakeholders.

Flexibility, adaptability, and learning are crucial. Flexibility calls for strategic adaptation in the wake of fluctuating macro environment factors. Examples include natural calamity, supply change dynamics, technological advancement, or changing regulations because of political shifts. Furthermore, strategic planning and research initiatives facilitate learning and knowledge in an ongoing manner. Within the context of heritage tourism, negotiated authentication processes should be harmoniously aligned with the core elements of resilience. Collectively, in the face of an externality, exploitation, or devastation triggered by a natural calamity such as a pandemic, they can inform and build relevant support mechanisms to assist in identifying damage and type and magnitude of intervention required to bring a heritage tourism system to an equilibrium state. For instance, non-pharmaceutical interventions (NPIs) (such as face masks, washing of hands, and social distancing) are currently the most effective defense response to COVID-19. Adaptive Heritage tourism programs and their ongoing authentication processes are taking the NPIs into consideration. Conservation and training initiatives and heritage experiences are being launched keeping the World Health Organization guidelines into consideration.

Moving forward, in the context of sustainable development of heritage tourism and its negotiated authentication in the resilience model, it can be argued that both complement each other as they advocate adaptiveness. The idea is to evolve and generate opportunities to restore equilibrium offset by a sudden chaotic event (COVID-19). Learning and knowledge inform of updates and can assist in organizing/rebuilding relevant socioecological and cultural systems when equilibrium is restored, for instance, with the development of an effective vaccine. The resilient systems have to be location specific so that they can adapt to the area peculiarities such as landscape, cultural values, and social/economic/physical/mental health of local communities. For instance, different countries are imposing pandemic restrictions based on their level of preparedness, resources, and threat. The resilience model posits that the system should have the capability to identify and build entrepreneurial prospects. Furthering the list, government intervention is also crucial for catastrophic events such as social regulations and economic relief to counter health risks and stabilize the collective well-being of the community at large. Time and time again, local communities have faced testing times and delicately adapted to remain afloat by uplifting local pride and sharing/offering cultural/heritage events; such events hold the capability to relax, edutain, and immerse audience which is seeking optimized authenticity and truth of the moment experiences.

To put it simply, Cochrane's 'Sphere of Tourism Resilience' is relevant to the contemporary upheaval, as an insightful guide, as it stretches the highly theoretical discussion of complexity and resilience to the realm of real life situations. It identifies factors crucial to make heritage tourism resilient and thus allows focus on supporting key elements of a dynamic system that are crucial to harmonize and open a stable pathway that can facilitate strategic authentication of negotiated authenticity premised on promoting heritage sustainability (Cochrane 2010).

In summary, Figure 12.1 captures the fundamental tenets of a resilient negotiated authentication system for heritage tourism. Through the proposed model, I argue that mutual and self-authentications, premised on a combined expression of existentialist (premised on virtuosity, distantiation, self-making, and self-recovery tools) and objective authenticities, are required to make a heritage tourism system resilient. The three-key authenticating role players are heritage agencies, traditional and/or marginalized communities, and tourists. An efficient and successful authentication system should be reliant on both immersive digital and digital detox innovations to promote experiential consumption by elevating positive emotions such as a strong sense of 'hereness' and belonging, satisfaction, de-stressing/relaxation, and happiness. In this context, cognitive appraisal theory (CAT) can offer deep insights on human emotions and behavior (Watson & Spence 2007). Cognitive appraisal helps determine relevance to one's well-being and refers to emotions as a mental state that is facilitated by evaluating personally relevant information (Roseman, Spindal, & Jose 1990). Based on literature review, Zheng, Ritchie, Benckendorff, and Bao identify three favorable emotions: joy, love, and positive surprise, which are identified "by appraisals of pleasantness, goal congruence and internal self-compatibility. Four appraisal dimensions are suggested namely outcome desirability, fairness, certainty and coping potential" (2019, p. 240). The goal of mutual and/or self-authentication should be to utilize these dimensions to optimize experiential consumption in an ethical/moral manner. In a restricted

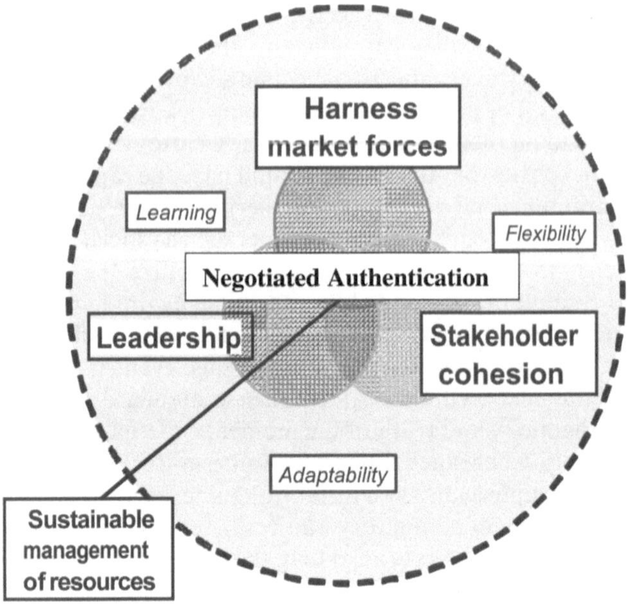

Figure 12.1 Sphere of tourism resilience.

Source: Cochrane (2010, p. 10)

manner, it can tap into the opportunities offered by immersive digital innovations such as digital surrogates and other multimodal applications and artificial intelligence (such as robotics).

Success of negotiated authentication, in the intra- and post-pandemic times is reliant on the resilience of the tourism system in the outer layer. As pinpointed earlier in this chapter strategic resilient framework needs to appropriately build and support a robust sustainable system in troubled times by facilitating: learning, flexibility, harnessing of market forces, stakeholder cohesion, effective leadership, and adaptability to travel restrictions. COVID-19 has offered a time for reflection and introspection. Several scholars view it as an opportunity to transform and reinvent tourism that is morally grounded, supports social justice, and is in harmony with all living beings.

Transformative times

Globalization coupled with advancement in technology had made our planet small with "porous borders and instantaneous knowledge sharing across the world has contributed to advances in the global economic well-being, enabling the outbreak of socially transmitted diseases" (like the COVID-19 coronavirus) "to be transmitted faster and across a broader geographic area than has ever been witnessed before" (Galvani, Lew, & Perez 2020). In fact, COVID-19 is an unequivocal outcome of globalization (Gössling Scott, & Hall 2020; Hall et al. 2020). The pandemic is still unfolding and it is risky to forecast the future. Some tourism destinations are struggling to start from the scratch. The current environment is calling for transformative solutions. The imprint of COVID-19 will stay for years to come. Destinations might come up with multiple pathways as they toil to attract different tourist markets. They might become unsustainable and less desirable by a new generation of morally self-authentic tourists or move toward delicately localized sustainable development but with less tourist demand. Transformational marketing strategies will need to be devised (Hall et al. 2020). A new level of global consciousness is likely to obtain a strong foothold in the long term and it is anticipated to transform ways of doing things and redefine the meaning of tourism. Some predictions of the new norm in the post-pandemic times include:

- Social distance may be here to stay, with periodic "suppress and lift" enforcements to control various disease outbreaks, along with an increased awareness of cleanliness and sanitation;
- A form of tourism 'degrowth' may emerge in which travel will be smaller and more costly, fostering fewer, but more meaningful and less hedonistic trips for most people, along with an increased appreciation of the 'right to travel';
- There may be an increased awareness of 'space', including global geographic space (what is happening on the opposite side of the world) and individual personal space (between people); related to this will be an increased appreciation for meaningful social interactions;
- There will likely be a major increase in the use of technologies and robotics to monitor people and products to ensure health and security for both, and to enhance communication and enable more work from home, educational opportunities, and other social interactions;
- There will likely be an increased appreciation and respect for the natural environment;
- Where possible, people will increase their home gardening, home cooking, and other home and self-care skills;
- Social enterprising will become more common as companies show more concern for workers and the surrounding community than for investor profits;
- There will be a general increase in awareness of, and support for, vulnerable populations, such as the homeless, informal economy workers, and refugees, as well as for frontline services workers, such as grocery store employees; and
- Shift might occur toward gender equity. Female political leaders will receive greater acknowledgment and respect following their apparent greater success in managing the COVID-19 pandemic incidents of

illness and death, and to keep their hospitals from overflowing with patients (Galvani et al. 2020, pp. 571–572). Galvani et al. further write that:

> COVID-19 has emerged as a new collective archetype, with deep emotional significance due to the global reach of its impacts. Thus, even if its long-term impacts on human behavior and related planetary outcomes may be challenging to predict and identify on a superficial conscious level, it is likely that its deeper, unconscious significances will be profound for generations to come. The smallest thing can affect the largest number of people, and what affects one can affect all. This is the lesson of the COVID-19 pandemic. (Galvani et al. 2020, p. 572)

Although distancing between people has increased, feelings of empathy and emotional solidarity exist and people have become close to each other, at the subconscious level. The covid virus has not made any distinction between the rich and the poor albeit the people with less privileges are likely to become more vulnerable and susceptible to infections. The entire world has a shared focus and health holds a new definition and meaning today. According to the World Health Organization: health is a state of complete physical, mental, and social well-being and not merely the absence of disease or infirmity; WHO's vision is of a world in which all peoples attain the highest possible level of health, and mission to promote health, keep the world safe, and serve the vulnerable, with measurable impact for people at country level (World Health Organization 2020).

UNESCO has taken several digital initiatives to continue their conservation work across several world heritage sites. It is also tapping into technology to promote local art and training of young heritage entrepreneurs. Trends are more likely to shift toward support of objectively or delicately negotiated authenticity of local heritage and of the self. And heritage tourism will continue but it is anticipated that it will occur in a more existentially authentic manner that promotes virtuosity and moral selving; this will elevate demand for more objectively authentic elements in heritage expressions and experiences but in a sustainable manner. Social distancing demands are likely to continue and impose a check on hedonistic and unsustainable growth of tourism. Heritage consumption, in fact tourism consumption in general, is likely to forge desire for the enhancement of moral selves and numerous consumer markets will make sincere efforts to embrace and demand ethical practices.

Almost simultaneously, the political processes are also likely to challenge the terms of ethical production, presentation, and consumption. Regimes in power have shaped the economic value of authenticity and legitimized it for preservation as a unique piece or property and they stand to benefit the most financially. For instance, in the context of authentic food systems, Cavanaugh writes that, "sustainability of heritage food requires it be sold by the true custodians who have the warrant to claim its ownership and lifestyle and culture is an embodiment of this production. They are themselves a part of the history" (2020, p. 3). Sameness has not been in demand and considered unexotic, hence controlled set of departures are planned while keeping the core essence intact. The author maps the manner in which:

> heritage is built on a foundation of authenticity elaborated with chronotopic bundlings of time and place, offering examples of smaller and large scale production for each and argues that constructing continuity across historical time – connecting 'now' to a particular and valued 'then' – is a semiotic imperative in the producing of heritage foods. Hands working in the past must be linked to hands working in the present… heritage is an issue or topic of contemporary concern. Time and place, then, must be bound up together in the labeling of heritage food. The chronotype of heritage connects modern tastes and desires to the practices and goods of a particular place and its past, a space-time fusion that adds value to the goods that can embody its qualisigns. (Cavanaugh 2019, pp. 5, 15)

Most of the emplaced chronotypes of heritage expressions are a product of institutional and autocratic interventions from government and non-government sectors. For transformations to happen, interventions should predominantly include local and marginalized voices to select/interpret/showcase authentic expressions of heritage. A resilient approach calls for ongoing innovation to prevent a system from dissolution and buried in the collapse stage. It needs to identify different layers of interventions and influence role players within each layer to optimize opportunities and resources (Cochrane 2010). Galvani et al. write that"COVID-19 is making us re-think how we conceive of ourselves as human beings, realizing that being human means caring……. and seeking harmony between the physical–emotional–mental–spiritual worlds. There are indications that a sustainable development empathy is emerging that will enhance our emotional and subconscious connections with peoples and environments" (2020, p. 573).

To put it simply, this pandemic is an uncommon occurrence and it has significantly disrupted the heritage tourism system and the way humans connect with cultural resources and seek authenticity outside and within themselves. Higher sense of global consciousness is a call to refine our authentic selves and rise above inauthenticity. We are at an introspective stage and facing a period of uncertainty; hence, at the current time of reckoning, all reorganization strategies have to revolve round social distancing, face coverings, and good hygiene. The reset point of this year is self-inquiry and awareness of dissonance within ourselves, so that moralized/virtuous self-authentication can take roots.

In closing, I would like to point out that the authenticity discourse will, undoubtedly, continue to evolve and scholars will innovatively pursue new trajectories and develop resilient schemas to make cultural/heritage resources sustainable, while optimizing self-making/self-recovery and self moralizing-authentication and self-gratification moments for the heritage audience (Figure 12.2). Authenticity and authentication of heritage tourism span a broad spectrum of discourses such as: host-guest authentication, knowledge transfer processes in heritage tourism, authenticity and anxiety in the smell of death and life, boundaries of authenticity, nostalgia, sustainability, smart marketing, destination competitiveness, self authentication, moral selving, authenticity value chains, authentic heritage tourism systems, affective connotations of authenticity, legitimization and deconstruction/reconstruction processes of different authenticities and, existence

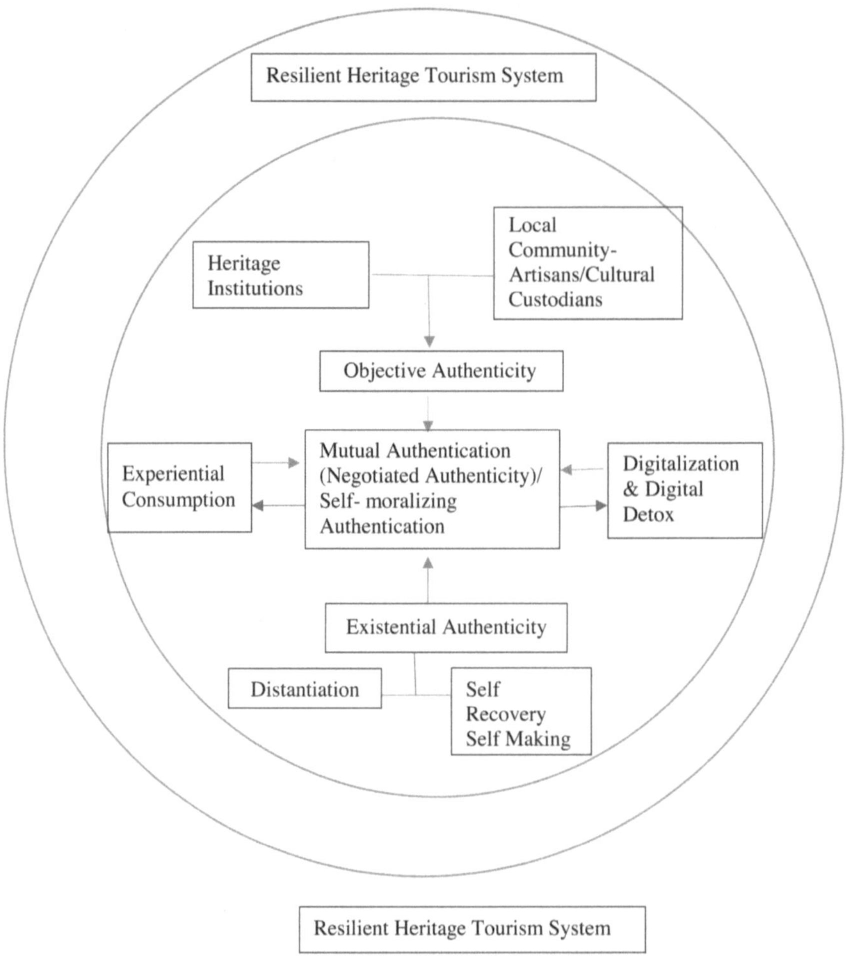

Figure 12.2 Resilient negotiated authentication system for heritage tourism in transformative times.

of power and authority interventions (Chhabra 2019). I have offered deliberations in many of the aforementioned areas, using an exploratory approach. Through this book, I suggest stimulating and creative routes to further a resilient authentication systems approach in heritage tourism.

I also point out that, despite the impressive scholarly progress in this field, academic dialogue on the politics of authentication and the impact of digital toxicity on different dimensions/experiences of authenticity remains scant. This topic can be furthered by examining the self-authentication process, the 'hereness' of digital detox moments, and "the role of tourists as authenticating agents of heritage, especially using digitalization as a backdrop and cases embracing co-created processes of resilient mutual authentication, between the tourists and heritage agencies/host communities" (Chhabra 2019, p. 6). Despite the much touted immersive benefits of digital surrogates and multimodal applications, a concern looms over how ICT is transforming dehumanized heritage tourism experiences and producing couch surfers and digital nomads. In the context of heritage tourism, one might ask: can technology adequately replace the 'human-face' of the authentic tangible and intangible experiences of heritage tourism? To what extent are heritage tourists willing to interact with robots when they seek interactive engagement with heritage resources and what is their perceived optimal trade-offs between dehumanized efficiency and personalized interactive experiences? What kind of negotiated authentication processes can optimize artificial intelligence in human settings?

As we approach the end of 2020, the development and diffusion of robotics and artificial intelligence in heritage tourism experiences and services ignite further deliberations, remarkably so in the context of sustainable use of ICT in authenticating and humanizing heritage tourism in a resilient manner. Furthermore, benefits of self-authentication and moral selving through cultural/heritage experiences remain unexplored, particularly in terms of quality of life, enhancement of the self from an eudaimonic perspective, mindfulness, and overall well-being of the society in the intra- and post-pandemic times. A transformative path in the field of authenticity and authentication necessitates advancement of future scholarship in this direction.

References

Anderson, A. R., & Knee, E. (2020). Queer isolation or queering isolation? Reflecting upon the Ramifications of COVID-19 on the Future of Queer Leisure Spaces. *Leisure Sciences*, 1–7.

Bartholomew, J. (2015). I invented "virtue signalling". Now it's taking over the world. *The spectator* (p. 10). Retrieved from https://www.spectator.co.uk/2015/ 10/i-invented-virtue-signalling-now-its-taking-over-the-world/ (Accessed September 09 2019).

Boaks, J., & Levine, M. P. (2015). *Leadership and ethics*. London: Bloomsbury Publishing.

Breathnach, T. (2006) Looking for the real me: Locating the self in heritage tourism. *Journal of Heritage Tourism*, *1*(2), 100–120.

Brown, L. (2013). Tourism: A catalyst for existential authenticity. *Annals of Tourism Research*, *40*(1), 176–190.

Cavanaugh, J. R. (2019). Labelling authenticity, or, how I almost got arrested in an Italian supermarket. *Semiotic Review*, (5).

Chhabra, D. (2008). Positioning museums on an authenticity continuum. *Annals of Tourism Research*, *35*(2), 427–447.

Chhabra, D. (2015). *Strategic marketing in hospitality and tourism*. Nova Science Publishers, Inc.

Chhabra, D. (2019). Authenticity and authentication: Dialogical perceptiveness. *Journal of Heritage Tourism*. In Press.

Cochrane, J. (2010). The sphere of tourism resilience. *Tourism Recreation Research*, *35*(2), 173–185.

Cohen, E. (1989). Primitive and remote: Hill tribe trekking in Thailand. *Annals of Tourism Research*, *16*(1), 30–61.

Cohen-Aharoni, Y. (2017). Guiding the 'real' temple: The construction of authenticity in heritage sites in a state of absence and distance. *Annals of Tourism Research*, *63*, 73–82.

Daigle, C. (2017). *Authenticity and distantiation from oneself: An ethico-political problem*. Project Muse: John Hopkins University Press.

Domínguez-Quintero, A., González-Rodríguez, M., & Roldán, J. (2018). The role of authenticity, experience quality, emotions, and satisfaction in a cultural heritage destination. *Journal of Heritage Tourism*, 1–15.

Ekelund, R. B., Higgins, R., & Jackson, J. D. (2019). ART as meta-credence: Authentication and the role of experts. *Journal of Cultural Economics*, 1–17.

Fu, Y., Liu, X., Wang, Y., & Chao, R. F. (2018). How experiential consumption moderates the effects of souvenir authenticity on behavioral intention through perceived value. *Tourism Management*, *69*, 356–367.

Galvani, A., Lew, A. A., & Perez, M. S. (2020). COVID-19 is expanding global consciousness and the sustainability of travel and tourism. *Tourism Geographies*, *22*(3), 567–576.

García-Esparza, J. (2016). Re-thinking the validity of the past. Deconstructing what authenticity and integrity mean to the fruition of cultural heritage. *VITRUVIO-International Journal of Architectural Technology and Sustainability*, *1*(1), 21–34.

Gino, F., Kouchaki, M., & Galinsky, A. D. (2015). The moral virtue of authenticity: How inauthenticity produces feelings of immorality and impurity. *Psychological Science*, *26*(7), 983–996.

Gossling, S., Scott, D., & Hall, M. (2020): Pandemics, tourism and global change: A rapid assessment of COVID-19, *Journal of Sustainable Tourism*, doi: 10.1080/09669582.2020.1758708.

Grau, C., Ginhoux, R., & Riera, A. (2003). Long term possibilities. *Defense & Technology Paper*, *57*.

Gunderson, L. (2000). Ecological resistance – in theory and application. *Annual Review of Ecological Systems*, *31*, 425–439.

Hall, C. M., Scott, D., & Gössling, S. (2020). Pandemics, transformations and tourism: Be careful what you wish for. *Tourism Geographies*, 1–22.

Hand, M. (2016). *Making digital cultures: Access, interactivity, and authenticity*. Routledge.

Heller, M., Pietikäinen, S., & da Silva, E. (2017). Body, nature, language: Artisans to artists in the commodification of authenticity. *Anthropologica*, *59*(1), 114–129.

Holling, C. (2001). Understanding the complexity of economic, ecological and social systems. *Ecosystems, 4*, 390–405.

Holroyd, A. T., Cassidy, T., Evans, M., Gifford, E., & Walker, S. (2015). Design for 'domestication': The decommercialisation of traditional crafts. In 11th European Academy of Design Conference, Boulogne Billancourt, France.

Hopper, D., Costley, C. L., & Friend, L. A. (2015). Embodied self-authentication. *Australasian Marketing Journal (AMJ), 23*(4), 319–324.

Jantz, R. C. (2009). An institutional framework for creating authentic digital objects.

Malzbender, T., Gelb, D., & Wolters, H. (2001). Polynomial Texture Maps. *Proceedings of ACM Siggraph 2001* MeshLab web, MeshLab website. Retrieved from http://meshlab.sourceforge.net (Accessed May 2007).

Markovic, K. (2018). *Authenticity in digital surrogates. Workflow development for generating an authentic digital surrogate for heritage conservation* (Doctoral dissertation, Auckland University of Technology).

McDermott, M., Schwartz, D., & Vallejo, S. (2015). Talking the talk but not walking the walk: Public reactions to hypocrisy in political scandal. *American Politics Research, 43*(6), 952–974.

Meng, Z., Cai, L. A., Day, J., Tang, C. H., Lu, Y., & Zhang, H. (2019). Authenticity and nostalgia – subjective well-being of Chinese rural-urban migrants. *Journal of Heritage Tourism*, 1–19.

Mkono, M. (2020). Eco-hypocrisy and inauthenticity: Criticisms and confessions of the eco-conscious tourist/traveller. *Annals of Tourism Research, 84*, 102967.

Mudge, M., Ashley, M., & Schroer, C. (2007, October). A digital future for cultural heritage. In *AntiCIPAting the Future of the Cultural Past, Proceedings of the XXI International CIPA Symposium* (pp. 1–6).

Phoenix Art Museum. (2020). Belief, Collection Highlights. Retrieved in August, 2020, from https://mailchi.mp/phxart/virtual_phxart_belief?e=59ac6d7477.

Pine, B., & Gilmore, J. H. (1999). *The experience economy: Work is theatre & every business a stage*. Harvard Business Press.

Poria, Y., Biran, A., & Reichel, A. (2006). Tourist perceptions: Personal vs. non-Personal. *Journal of Heritage Tourism, 1*(2), 121–132.

Reisinger, Y., & Steiner, C. (2006). Reconceptualising interpretation: The role of tour guides in authentic tourism. *Current Issues in Tourism, 9*(6), 481–498.

Rinder, I. D., & Campbell, D. T. (1952). Varieties of inauthenticity. *Phylon (1940–1956), 13*(4), 270–275.

Roseman, I., Spindal, M., & Jose, P. (1990). Appraisals of emotion-eliciting events: Testing a theory of discrete emotions (attitudes and social cognition). *Journal of Personality and Social Psychology, 59*(5), 899.

Song, H., Liu, J., & Chen, G. (2012). Tourism value Chain governance: Review and prospects. *Journal of Travel Research, 20*(10), 94–103.

Stanco, F., Battiato, S., & Gallo, G. (Eds.). (2011). *Digital imaging for cultural heritage preservation: Analysis, restoration, and reconstruction of ancient artworks*. CRC Press.

Stankard, S. (2010). *Textile Praxis: The Case for Malaysian Hand-Woven Songket*. Unpublished PhD thesis. Royal College of Art.

Steiner, C. J., & Reisinger, Y. (2006). Understanding existential authenticity. *Annals of Tourism Research, 33*(2), 299–318.

Stevenson, N., Airey, D., & Miller, G. (2009). Complexity Theory and Tourism Policy Research. Tourism Research (Paper 54). Retrieved from http://epubs.surrey.ac.uk/tourism/54 (Accessed May 2019).

Syvertsen, T., & Enli, G. (2019). Digital detox: Media resistance and the promise of authenticity. *Convergence: The International Journal of Research into Modern Media Technologies*, 1–15.

Trivers Theory. (1985). The evolution of reciprocal altruism. *Quarterly Review of Biology*, *46*, 35–57.

Trivers Theory. (2006). Reciprocal altruism: 30 years later. In *Cooperation in primates and humans* (pp. 67–83). Berlin, Heidelberg: Springer.

UNESCO (2003). Guidelines for the Preservation of Digital Heritage. Retrieved May, 2020, from http://www.unesco.org/new/en/communication-and-information/resources/publications-and-communication-materials/publications/full-list/guidelines-for-the-preservation-of-digital-heritage/.

Wang, N. (1999). Rethinking authenticity in tourism experience. *Annals of Tourism Research*, 26(2), 349–370.

Watson, I., & Spence, M. (2007). Causes and consequences of emotions on consumer behavior: A review and integrative cognitive appraisal theory. *European Journal of Marketing*, *41*(5/6), 487–511.

Weisstouch, L. (2020). In defense of souvenirs: For travel-lovers stuck at home, even the unlikeliest tchotchkes can spark joy. Retrieved June, 2020, from https://www.washingtonpost.com/life(style/travel/in-defense-of-souvenirs-to-travel-lovers-stuck-at-home-even-chintzy-objects-can-spark-joy/2020/04/16/a281783a-7da6-11ea-9040-68981f488eed_story.html.

World Health Organization (2020). Vision and highlights. Retrieved from https://www.who.int/

Zheng, D., Ritchie, B. W., Benckendorff, P. J., & Bao, J. (2019). Emotional responses toward tourism performing arts development: A comparison of urban and rural residents in China. *Tourism Management*, *70*, 238–249.

Index

Note: Page numbers in *italics* indicate figures and **bold** indicates tables in the text.